Nonlinear Cosmic Ray Diffusion Theories

Nonlinear Cosmic Ray Diffusion Theories

by

Andreas Shalchi

Springer

Andreas Shalchi
Universität Bochum
Lehrstuhl für Theoretische Physik IV
Universitätsstr. 150
44780 Bochum
Germany

Cover image: Credit by NASA/CXC/Rutgers/J. Warren & J. Hughes et al.
© by A. Shalchi

ISSN: 0067-0057
ISBN: 978-3-642-10139-7 e-ISBN: 978-3-642-00309-7
DOI: 10.1007/978-3-642-00309-7

Cover design: eStudio Calamar S.L.

Printed on acid-free paper

9 8 7 6 5 4 3 2 1

Springer is part of Springer Science+Business Media (www.springer.com)

Preface

If charged particles move through the interplanetary or interstellar medium, they interact with a large-scale magnetic field such as the magnetic field of the Sun or the Galactic magnetic field. As these background fields are usually nearly constant in time and space, they can be approximated by a homogeneous field. If there are no additional fields, the particle trajectory is a perfect helix along which the particle moves at a constant speed. In reality, however, there are turbulent electric and magnetic fields due to the interstellar or solar wind plasma. These fields lead to scattering of the cosmic rays parallel and perpendicular to the background field. These scattering effects, which usually are of diffusive nature, can be described by spatial diffusion coefficients or, alternatively, by mean free paths. The knowledge of these parameters is essential for describing cosmic ray propagation as well as diffusive shock acceleration. The latter process is responsible for the high cosmic ray energies that have been observed.

The layout of this book is as follows. In Chap. 1, the general physical scenario is presented. We discuss fundamental processes such as cosmic ray propagation and acceleration in different systems such as the solar system or the interstellar space. These processes are a consequence of the interaction between charged cosmic particles and an astrophysical plasma (turbulence). The properties of such plasmas are therefore the subject of Chap. 2. The standard approach for describing plasma–particle interactions is a first order perturbation theory, also known as quasilinear theory. This theory, which was historicaly the first and most applied theory, is reviewed in Chap. 3. By using computer simulations for investigating plasma–particle interactions, it has been discovered that, for several cases, quasilinear theory is inaccurate. Therefore, several nonlinear theories for describing the motion of charged particles through an astrophysical plasma have been developed in the past. The most prominant nonlinear approaches are presented in Chaps. 4–7. In Chap. 8, these theories are applied on different physical situations such as cosmic ray propagation in the solar system and in the interstellar medium. Furthermore, we revisit the problem of diffusive shock acceleration. The latter mechanism is responsible for the high cosmic ray energies and depends strongly on the particle diffusion coefficient that has to be calculated nonlinearly. In the concluding Chap. 9, we summarize the results of this book and give an outlook about the future work on the field of cosmic ray diffusion theory.

I am grateful to many colleages, friends, and students for their support and encouragements. I am especially grateful to Reinhard Schlickeiser, from whom I learned a lot about the physics of cosmic rays. I am thankful to John W. Bieber and William H. Matthaeus for enabling me to visit the Bartol Research Institute of the University of Delaware as a Post Doctoral Fellow.

Moreover, I want to thank William H. Matthaeus, Jaco Minnie, and Gang Qin for providing the test particle code that was used to perform the simulations shown in this book. I also thank Gary P. Zank for providing some of the illustrations and for the nice time that I have spend as a visiting scientist at the Center for Space Plasma and Aeronomic Research (CSPAR) located at the University of Alabama, Huntsville.

Special thanks go to Alexander Dosch, Horst Fichtner, Ioannis Kourakis, Robert Tautz, and Bastian Weinhorst for carefully reading parts of the manuscript and for their valuable suggestions. Needless to say, all remaining errors are my own. This book would not have been written without the support of my wife Shabnam and my son Richard.

Bochum, May 2009 *Andreas Shalchi*

Contents

Acronyms

AMPTE	Active magnetospheric particle tracer explorer
BAM model	Bieber and Matthaeus model
CME	Coronal mass ejection
DT model	Damping model of dynamical turbulence
ENLGC theory	Extended nonlinear guiding center theory
ESP	Energetic storm particle
FL	Field line
FLRW	Field line random walk
ISM	Interstellar medium
LTA	Late time approximation
MHD	Magnetohydrodynamics
MSD	Mean square deviation
NADT model	Nonlinear anisotropic dynamical turbulence model
NCA	Nonlinear closure approximation
NLGC theory	Nonlinear guiding center theory
NPT	Nonlinear perturbation theory
PAFT	Partially averaged field theory
PW model	Plasma wave model
QLT	Quasilinear theory
RS model	Random sweeping model
SEP	Solar energetic particle
SOQLT	Second-order quasilinear theory
SQLT	Standard quasilinear theory
TGK formulation	Tylor–Green–Kubo formulation
WNLT	Weakly nonlinear theory

Chapter 1
The General Scenario

Within cosmic ray transport theory, we investigate the interaction between energetic charged particles like electrons, protons, or heavy ions and astrophysical plasmas such as the solar wind or the interstellar medium. These particles interact with a background magnetic field $\mathbf{B}_0$ and with turbulent electric and magnetic fields $\delta\mathbf{E}$ and $\delta\mathbf{B}$, and they therefore experience scattering parallel and perpendicular to $\mathbf{B}_0$. In this introductory chapter, general properties of cosmic rays are discussed, as well as the unperturbed motion of the particles. Furthermore, the physics of parallel and perpendicular scattering is investigated. At the end of this chapter, we consider observed mean free paths of cosmic rays in the heliosphere and in the interstellar medium. One aim of this book is to demonstrate that a nonlinear description of particle transport is necessary to reproduce these measurements.

1.1 Cosmic Rays

In the beginning of the twentieth century, physicists were unable to explain the source of the ionizing radiation that penetrated through electroscopes aboard balloons. For a long time it was assumed that the radiation must have originated from a source on Earth. In several balloon flights between the years 1911 and 1913, the Austrian scientist Victor F. Hess discovered that the radiation increased rapidly with altitude. As a consequence, he concluded that the radiation is extraterrestrial in origin. In the year 1925, Robert A. Millikan confirmed Hess's theory and introduced the name *cosmic rays* for this radiation.

In the following years, cosmic ray research became an interesting and important part of physics. A prominent example is the discovery of several new particles, including the positron. Furthermore, cosmic ray research led to huge advances in astrophysics, space physics, and cosmology. As a consequence, Victor F. Hess received the Nobel Prize for Physics in the year 1936 for his discovery of cosmic rays.

A. Shalchi, *Nonlinear Cosmic Ray Diffusion Theories*,
Astrophysics and Space Science Library,
DOI: 10.1007/978-3-642-00309-7_1, © Springer-Verlag Berlin Heidelberg 2009

1.1.1 General Properties of Cosmic Rays[1]

The energy of cosmic rays is usually measured in units of megaelectron volts (MeV) or gigaelectron volts (GeV). The highest cosmic ray energy that has been measured to date was more than 10^{20} eV. Some particles are from sources located in the solar system (solar cosmic rays), while other cosmic rays originate from extrasolar sources within our own Galaxy, such as the supernova remnants. However, the fact that some cosmic rays have extremely high energies provides evidence that at least some of them must be of extragalactic in origin; the local galactic magnetic field would not be able to contain particles with such high energy. The origin of cosmic rays with energies up to 10^{15} eV can be accounted for in terms of shock-wave acceleration in supernova shells. Recent observations of the *Pierre Auger Collaboration* found some evidence that cosmic rays with energies greater than 10^{15} eV originated from *active galactic nuclei (AGN)* (see The Pierre Auger Collaboration 2007). This evidence is also in agreement with some theoretical work (see Honda and Honda 2004). These authors have shown that particles can get accelerated in AGN shock waves up to 10^{21} eV.

Observations have shown that cosmic rays with an energy above 10 GeV approach the Earth's surface isotropically. It has been hypothesized that this is not due to an even distribution of cosmic ray sources, but instead is due to galactic magnetic fields causing cosmic rays to travel in spiral paths (see e.g., Figs. 1.1 and 1.2). At energies below 10 GeV, there is a directional dependence due to the interaction of the charged components of the cosmic rays with the Earth's magnetic field.

Depending on the origin of the cosmic ray components, we usually distinguish between at least four different cosmic ray species (see Table 1.1): solar cosmic rays, anomalous cosmic rays, galactic cosmic rays, extragalactic/ultra-high energy cosmic rays. Besides the question of the origin of cosmic rays, a main problem is the interaction between the charged cosmic particles with the turbulent electromagnetic fields of the interstellar or interplanetary medium. Because cosmic rays

Table 1.1 The cosmic ray population

Cosmic ray species	Energy/nucleon
Solar cosmic rays	10^{1}–10^{9} eV
Anomalous cosmic rays	10^{6} eV
Galactic cosmic rays	10^{9}–10^{15} eV
Extragalactic/ultra-high energetic cosmic rays	10^{15}–10^{20} eV

Shown are the typical energies of the cosmic ray species in electron volts (eV)

[1] More detailed information about cosmic ray properties can be found in Fichtner (2000, 2001) or Schlickeiser (2002). More introductory information about cosmic rays can be found in *http://en.wikipedia.org*, *http://www.britannica.com*, and an article of R.A. Mewaldt, which can be found online at *http://www.srl.caltech.edu/personnel/dick/cos_encyc.html*

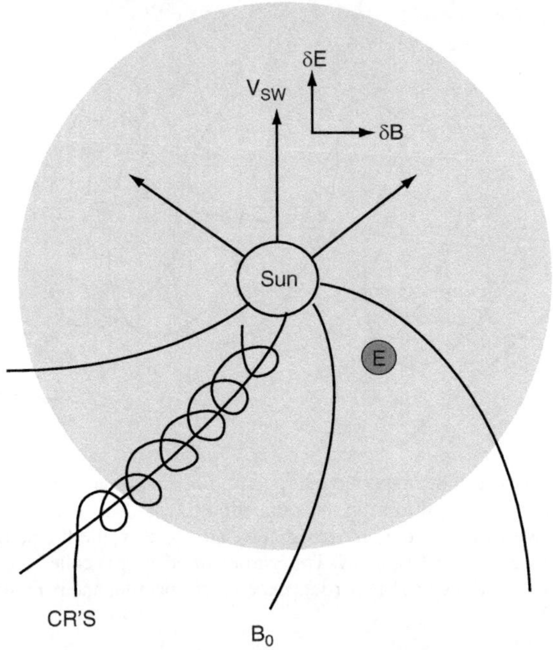

Fig. 1.1 The heliosphere is a prominent example to study the transport of cosmic rays (CRs). The charged particles interact with the magnetic field of the Sun, which is approximately constant at the time and length scales of interest, and with the turbulence due to the solar wind plasma

are electrically charged, they are deflected by magnetic fields and their directions are randomized, making it impossible to tell where they originated from. However, cosmic rays in other regions of the Galaxy can be traced by their electromagnetic radiation. Supernova remnants such as the Crab Nebula are known to be a source of cosmic rays from the radio synchrotron radiation emitted by cosmic ray electrons spiraling in the magnetic fields of the remnant. In addition, observations of high-energy gamma rays resulting from cosmic ray collisions with interstellar gas show that most cosmic rays are confined to the disk of the Galaxy, presumably by its magnetic field.

If the turbulent fields due to the interstellar or interplanetary plasma were absent, the particles would follow the magnetic field of the Sun or the interstellar magnetic field by performing a helical motion, as shown in Figs. 1.1 and 1.2. The turbulent fields affect the trajectory of the particles in two ways:

1. The particles experience parallel and perpendicular scattering with respect to the background field (mainly due to the turbulent magnetic fields). In most cases, we expect to find a diffusive motion of the charged particles.
2. The particles experience momentum diffusion or stochastic acceleration (due to the turbulent electric fields).

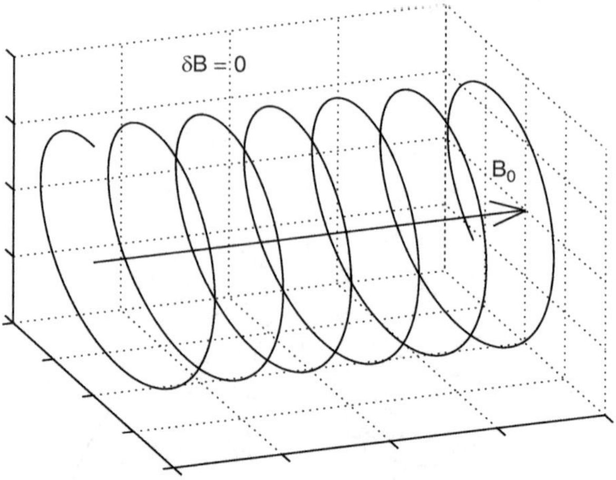

Fig. 1.2 Particle transport in the unperturbed system: if the turbulent fields are absent ($\delta B_i = 0$) and if the background field is uniform in the z-direction ($\mathbf{B}_0 = B_0\mathbf{e}_z$), the particle trajectory is a perfect helix along which the particle moves at constant speed, while the perpendicular velocity components oscillate at angular frequency. This particle trajectory is called the unperturbed orbit. In reality, when there is turbulence, the particle experiences parallel and perpendicular scattering

These scattering effects are usually described by diffusion coefficients. The knowledge of these parameters is essential for the following:

- Exploring particle transport in the solar corona (e.g., Fletcher 1997)
- Investigating solar energetic particles (e.g., Dröge 2000)
- Describing the modulation of galactic cosmic rays (e.g., Burger and Hattingh 1998)
- Calculating the lifetime of cosmic rays in the Galaxy (e.g., Jokipii and Parker 1969; Ptuskin 2001)
- Describing shock acceleration of particles at parallel and perpendicular interplanetary shock waves (e.g., Zank et al. 2006)
- Exploring acceleration of charged particles in supernova remnants (e.g., Berezhko et al. 2002; Berezhko and Völk 2007)
- Explaining the origin of ultra-high energetic cosmic rays (e.g., Honda and Honda 2004)

1.1.2 Cosmic Rays in the Solar System

Apart from being deflected by the magnetic fields in interstellar space, cosmic rays are also affected by the interplanetary magnetic fields due to the solar wind plasma, and have, therefore, difficulty in reaching the inner solar system. Spacecraft venturing out towards the boundary of the solar system have found that the intensity of galactic cosmic rays increases with the distance from the Sun. As solar activity varies over the 11-year solar cycle, the intensity of cosmic rays on Earth also varies, in anticorrelation with the sunspot number.

The Sun is also a sporadic source of cosmic ray nuclei and electrons that are accelerated by shock waves traveling through the corona, and by the magnetic energy released in solar flares. During such events, the intensity of energetic particles in space can increase by a factor of 10^2–10^6 from hours to days. Such solar particle events are much more frequent during the active phase of the solar cycle. The maximum energy reached in solar particle events is typically 10–100 MeV, occasionally reaching 1 GeV (roughly once a year) to 10 GeV (roughly once a decade). Solar energetic particles can be used to measure the elemental and isotopic composition of the Sun, thereby complementing spectroscopic studies of solar material.

A further component of cosmic rays, comprised only of those elements that are difficult to ionize, including He, N, O, Ne, and Ar, was given the name *anomalous cosmic rays* because of its unusual composition. Anomalous cosmic rays originate from electrically neutral interstellar particles that have entered the solar system unaffected by the magnetic field of the solar wind, been ionized, and then accelerated at the shock wave, formed when the solar wind slows as a result of plowing into the interstellar gas, presently thought to occur somewhere between 75 and 100 AU from the Sun.

In Fig. 1.1, it is illustrated how an energetic cosmic ray particle moves through the heliosphere. The particle trajectory is approximately a helix along the mean field, which, in this case, is the magnetic field of the Sun. As a very crude approximation, one could neglect the turbulent fields due to the solar wind plasma ($\delta B_i = 0$, $\delta E_i = 0$). The resulting system, which is usually called the unperturbed system, is discussed in the next paragraph. In reality, however, the turbulent fields are not absent and give rise to a perturbed motion of the particle. With time, the particle trajectory deviates more and more from the unperturbed motion. After a characteristic time-scale, the particle begins to move diffusively parallel and perpendicular to the background magnetic field. Therefore, the propagation of charged particles can be described by a diffusion equation. The theoretical and observational determination of diffusion coefficients is, therefore, an essential part of cosmic ray physics. It is the purpose of this book to describe new possibilities to calculate diffusion coefficients nonlinearly and to compare them with test-particle simulations and observations. In this introductory chapter of this book, the physics of parallel and perpendicular spatial diffusion is discussed (see Sects. 1.4 and 1.5).

1.2 The Unperturbed System

The motion of charged particles in electromagnetic fields is described by the Newton–Lorentz equation

$$\frac{\mathrm{d}}{\mathrm{d}t}\mathbf{p} = q\left(\mathbf{E} + \frac{\mathbf{v}}{c} \times \mathbf{B}\right). \qquad (1.1)$$

Here, the particle charge (q), the electric and magnetic fields ($\mathbf{E}, \mathbf{B}$), the particle velocity ($\mathbf{v}$), the particle momentum ($\mathbf{p}$), and the speed of light (c) are used. We choose our Cartesian system of coordinates so that the z-axis is aligned parallel to the mean field (background field) $\mathbf{B}_0$. Furthermore, we approximate the absolute value of the mean magnetic field B_0 by a constant field. Consequently, we have

$$\langle \mathbf{B} \rangle = \mathbf{B}_0 = B_0 \mathbf{e}_z. \tag{1.2}$$

In the case of heliospheric particle propagation, the mean field can be identified by the magnetic field of the Sun. Because of the high conductivity of cosmic plasmas, there are no large-scale electric fields

$$\langle \mathbf{E} \rangle = \mathbf{E}_0 = 0, \tag{1.3}$$

and we thus have

$$\mathbf{B} = B_0 \mathbf{e}_z + \delta \mathbf{B}, \quad \mathbf{E} = \delta \mathbf{E} \tag{1.4}$$

with the turbulent electric and magnetic fields ($\delta\mathbf{E}$, $\delta\mathbf{B}$). In this book, only turbulence models without electric fields ($\delta E_i = 0$) are considered. The main reason for using the model of purely magnetic fluctuations is that the electric fields are much smaller than the magnetic fields (see Sect. 2.5). Therefore, electric fields are less important for spatial diffusion. In purely magnetic systems ($\delta\mathbf{E} = 0$), we have

$$\mathbf{p} \cdot \dot{\mathbf{p}} = q \mathbf{p} \cdot \mathbf{E} = 0 \Rightarrow \dot{\mathbf{p}} = 0, \tag{1.5}$$

and thus the kinetic energy of the particle is conserved. Hence, we have $v = \text{const}$, the equations of motion can be written as

$$\begin{aligned}
\dot{v}_x &= \Omega v_y + \Omega \left(v_y \frac{\delta B_z}{B_0} - v_z \frac{\delta B_y}{B_0} \right), \\
\dot{v}_y &= -\Omega v_x + \Omega \left(v_z \frac{\delta B_x}{B_0} - v_x \frac{\delta B_z}{B_0} \right), \\
\dot{v}_z &= \Omega \left(v_x \frac{\delta B_y}{B_0} - v_y \frac{\delta B_x}{B_0} \right),
\end{aligned} \tag{1.6}$$

where we used

$$p_i = m \frac{v_i}{\sqrt{1 - v^2/c^2}} \tag{1.7}$$

and the parameter

$$\Omega := \frac{q B_0}{mc} \sqrt{1 - v^2/c^2}. \tag{1.8}$$

For the unperturbed system ($\delta B_i = 0$), the equations of motion are

$$\begin{aligned}\dot{v}_x &= \Omega v_y,\\ \dot{v}_y &= -\Omega v_x,\\ \dot{v}_z &= 0.\end{aligned} \tag{1.9}$$

These equations can easily be solved by

$$\begin{aligned}v_x &= v_\perp \cos(\Phi_0 - \Omega t),\\ v_y &= v_\perp \sin(\Phi_0 - \Omega t),\\ v_z &= v_\parallel = \text{const},\end{aligned} \tag{1.10}$$

where we used the velocity components parallel ($v_\parallel$) and perpendicular ($v_\perp$) to the background field $\mathbf{B}_0$. The parameter Φ_0 denotes the initial gyrophase. For the particle trajectory, we therefore find

$$\begin{aligned}x(t) &= x(0) + \frac{v_\perp}{\Omega}\sin(\Phi_0) - \frac{v_\perp}{\Omega}\sin(\Phi_0 - \Omega t),\\ y(t) &= y(0) - \frac{v_\perp}{\Omega}\cos(\Phi_0) + \frac{v_\perp}{\Omega}\cos(\Phi_0 - \Omega t),\\ z(t) &= z(0) + v_\parallel t.\end{aligned} \tag{1.11}$$

Obviously, we find a gyromotion (rotation) of the particle (see Fig. 1.2). The center of the rotation in the x-y plane is

$$\begin{aligned}x_m &= x(0) + \frac{v_\perp}{\Omega}\sin(\Phi_0),\\ y_m &= y(0) - \frac{v_\perp}{\Omega}\cos(\Phi_0),\end{aligned} \tag{1.12}$$

and the radius (gyroradius) is

$$R_\mathrm{G} = \frac{v_\perp}{\Omega}. \tag{1.13}$$

The parameter Ω introduced in (1.8) can easily be identified with the gyrofrequency of the unperturbed particle. It is convenient to introduce the so-called pitch-angle ϕ_P, which is the angle between the background field and the particle velocity vector. In the unperturbed system, the pitch-angle is constant and thus the pitch-angle cosine $\mu = \cos(\phi_\mathrm{P})$ is conserved. Then the parallel and perpendicular velocity components can be written as

$$v_\parallel = v\mu, \quad v_\perp = v\sqrt{1-\mu^2} \tag{1.14}$$

and the gyroradius is

$$R_\mathrm{G}(\mu) = \frac{v_\perp}{\Omega} = \frac{v}{\Omega}\sqrt{1-\mu^2} = R_\mathrm{L}\sqrt{1-\mu^2}, \tag{1.15}$$

where we used the Larmor radius $R_\mathrm{L} = v/\Omega = \text{const}$.

1.3 Particle Diffusion and the TGK Formulation

More challenging than the unperturbed system is to study particle propagation in a turbulence. In this case, the particles experience scattering parallel and perpendicular to the background magnetic field. In Sect. 1.4, we discuss in detail the physics of parallel diffusion, and in Sect. 1.5, the physics of perpendicular scattering. Here we present some general considerations. In the following, we use the variable x, which can be substituted by any component of the Cartesian coordinates.

1.3.1 Mean Square Displacements and Diffusion Coefficients

The mean square displacement is defined as

$$\left\langle (\Delta x)^2 \right\rangle = \left\langle (x(t) - x(0))^2 \right\rangle, \tag{1.16}$$

where we introduced the averaging operator $\langle \ldots \rangle$. By assuming

$$\left\langle (\Delta x)^2 \right\rangle \sim t^{\sigma} \tag{1.17}$$

for the temporal behavior of the mean square displacement, we can characterize the particle motion by using different parameter regimes of σ:

$$\begin{aligned} 0 < \sigma < 1 &: \text{subdiffusion,} \\ \sigma = 1 &: \text{normal (Markovian) diffusion,} \\ 1 < \sigma < 2 &: \text{superdiffusion,} \\ \sigma = 2 &: \text{ballistic motion (free streaming).} \end{aligned} \tag{1.18}$$

Cases with $\sigma > 2$ are not known in cosmic rays transport theory and are, therefore, not discussed in this book. In most cases, particle transport in astrophysical turbulence behaves diffusively ($\sigma = 1$). Only a few cases are known for which particle transport behaves sub- or superdiffusively. We define a diffusion coefficient κ_{xx} as

$$\kappa_{xx} = \lim_{t \to \infty} \frac{\left\langle (\Delta x)^2 \right\rangle}{2t}. \tag{1.19}$$

It should be noted, however, that in reality it is not possible to consider infinitely late times because of the finite size of the system. Then, the condition $t \to \infty$ has to be replaced by $t \gg t_{\mathrm{d}}$, where t_{d} is a characteristic time-scale that the particle needs in order to reach diffusive behavior. Furthermore, one can define a time-dependent or *running diffusion coefficient* as

$$\kappa_{xx}(t) = \frac{\left\langle (\Delta x)^2 \right\rangle}{2t} \tag{1.20}$$

or, as

$$d_{xx}(t) = \frac{1}{2}\frac{\mathrm{d}}{\mathrm{d}t}\left\langle(\Delta x)^2\right\rangle \equiv \frac{\mathrm{d}}{\mathrm{d}t}\left(t\kappa_{xx}(t)\right). \tag{1.21}$$

In the case of diffusion ($\kappa_{xx}(t \to \infty) \to \kappa_{xx} = \text{const}$), both definitions are equal. In principle, (1.19) can also be used to calculate a diffusion coefficient, but often it is more convenient to apply the so-called TGK formulation, which is discussed in the following paragraph.

1.3.2 The TGK Formulation

The Taylor–Green–Kubo formulation (TGK formulation, e.g., Taylor 1922; Green 1951; Kubo 1957) is a useful tool to calculate diffusion coefficients. In general, one can write for the mean square displacement

$$\begin{aligned}\left\langle(\Delta x)^2\right\rangle(t) &= \left\langle\left(\int_0^t \mathrm{d}\tau v_x(\tau)\right)^2\right\rangle,\\ &= \int_0^t \mathrm{d}\tau \int_0^t \mathrm{d}\xi\ \langle v_x(\tau)v_x(\xi)\rangle,\\ &= \int_0^t \mathrm{d}\tau \int_0^\tau \mathrm{d}\xi\ \langle v_x(\tau)v_x(\xi)\rangle\\ &\quad + \int_0^t \mathrm{d}\tau \int_\tau^t \mathrm{d}\xi\ \langle v_x(\tau)v_x(\xi)\rangle.\end{aligned} \tag{1.22}$$

To proceed, we assume homogenity in time and, therefore, that the velocity correlation functions depend only on the time difference, for example,

$$\langle v_x(\tau)v_x(\xi)\rangle = \langle v_x(\tau-\xi)v_x(0)\rangle \tag{1.23}$$

in the first integral, and

$$\langle v_x(\tau)v_x(\xi)\rangle = \langle v_x(\xi-\tau)v_x(0)\rangle \tag{1.24}$$

in the second one. Thus, we have

$$\begin{aligned}\left\langle(\Delta x)^2\right\rangle(t) &= \int_0^t \mathrm{d}\tau \int_0^\tau \mathrm{d}\xi\ \langle v_x(\tau-\xi)v_x(0)\rangle\\ &\quad + \int_0^t \mathrm{d}\tau \int_\tau^t \mathrm{d}\xi\ \langle v_x(\xi-\tau)v_x(0)\rangle.\end{aligned} \tag{1.25}$$

By using the integral transformation $\tau - \xi \to \xi$ in the first and $\xi - \tau \to \xi$ in the second integral, we find

$$\left\langle (\Delta x)^2 \right\rangle (t) = \int_0^t \mathrm{d}\tau \int_0^\tau \mathrm{d}\xi \, \langle v_x(\xi) v_x(0) \rangle + \int_0^t \mathrm{d}\tau \int_0^{t-\tau} \mathrm{d}\xi \, \langle v_x(\xi) v_x(0) \rangle \,. \quad (1.26)$$

By inserting $1 = \mathrm{d}\tau / \mathrm{d}\tau$ in both integrals and by using partial integration (p.I.), we have

$$\begin{aligned} \left\langle (\Delta x)^2 \right\rangle (t) &= \int_0^t \mathrm{d}\tau \frac{\mathrm{d}\tau}{\mathrm{d}\tau} \int_0^\tau \mathrm{d}\xi \, \langle v_x(\xi) v_x(0) \rangle \\ &+ \int_0^t \mathrm{d}\tau \frac{\mathrm{d}\tau}{\mathrm{d}\tau} \int_0^{t-\tau} \mathrm{d}\xi \, \langle v_x(\xi) v_x(0) \rangle, \\ &\overset{p.I.}{=} \int_0^t \mathrm{d}\tau (t-\tau) \, \langle v_x(\tau) v_x(0) \rangle + \int_0^t \mathrm{d}\tau \, \tau \, \langle v_x(t-\tau) v_x(0) \rangle, \\ &= 2 \int_0^t \mathrm{d}\tau (t-\tau) \, \langle v_x(\tau) v_x(0) \rangle. \end{aligned} \quad (1.27)$$

From this result, the running diffusion coefficient $d_{xx}(t)$ can be derived

$$\begin{aligned} d_{xx}(t) &= \frac{1}{2} \frac{\mathrm{d}}{\mathrm{d}t} \left\langle (\Delta x)^2 \right\rangle (t), \\ &= \frac{1}{2} \frac{\mathrm{d}}{\mathrm{d}t} 2 \int_0^t \mathrm{d}\tau (t-\tau) \, \langle v_x(\tau) v_x(0) \rangle, \\ &= \int_0^t \mathrm{d}\tau \, \langle v_x(\tau) v_x(0) \rangle \,. \end{aligned} \quad (1.28)$$

If particle transport is diffusive, we have by definition $d_{xx}(t \to \infty) \to \kappa_{xx} = \text{const.}$ Therefore, we can extend the upper limit of the integral in (1.28) to infinity to find

$$\kappa_{xx} = \int_0^\infty \mathrm{d}\tau \, \langle v_x(\tau) v_x(0) \rangle. \quad (1.29)$$

Equation (1.29) is the Kubo formula for the diffusion coefficient κ_{xx}. Furthermore, a simple relation for the velocity correlation function can be deduced from (1.28)

$$\frac{\mathrm{d}}{\mathrm{d}t} d_{xx}(t) = \frac{\mathrm{d}}{\mathrm{d}t} \int_0^t \mathrm{d}\tau \, \langle v_x(\tau) v_x(0) \rangle = \langle v_x(t) v_x(0) \rangle =: V_{xx}(t). \quad (1.30)$$

This relation is very useful to determine the velocity correlation function analytically (see e.g., Shalchi and Döring 2007) and numerically from test-particle simulations, where the function $d_{xx}(t)$ can be *measured* (see e.g., Qin et al. 2002a, b, 2006).

1.4 The Physics of Parallel Scattering

Because of the turbulent fields (δB_i), the cosmic ray particles experience scattering parallel to the background field. Here, it is demonstrated that pitch-angle diffusion in phase-space leads to parallel spatial diffusion in real space. In Fig. 1.3, a typical particle, which experiences parallel scattering, is shown.

1.4.1 The Two-Dimensional Fokker–Planck Equation

The basis to study parallel transport of cosmic rays is the two-dimensional Fokker–Planck equation

$$\frac{\partial f}{\partial t} + v\mu \frac{\partial f}{\partial z} = \frac{\partial}{\partial \mu}\left(D_{\mu\mu} \frac{\partial f}{\partial \mu} \right), \tag{1.31}$$

with the pitch-angle Fokker–Planck coefficient

$$D_{\mu\mu} = \int_0^{\infty} \mathrm{d}t \, \langle \dot{\mu}(t) \dot{\mu}(0) \rangle. \tag{1.32}$$

This formula can be derived systematically from the relativistic Vlasov equation. Schlickeiser (2002) derived the full six-dimensional Fokker–Planck equation. By assuming smallness of the perpendicular Fokker–Planck coefficients and by

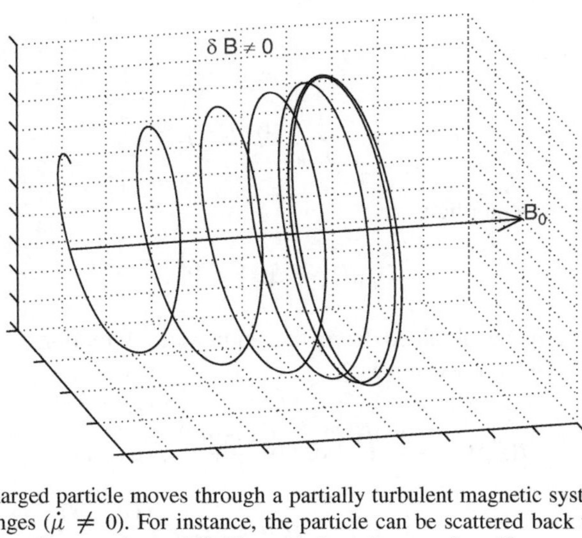

Fig. 1.3 If a charged particle moves through a partially turbulent magnetic system, the particle's pitch-angle changes ($\dot{\mu} \neq 0$). For instance, the particle can be scattered back to its initial position due to pitch-angle scattering at 90°. These pitch-angle scattering effects cause parallel spatial diffusion

neglecting momentum diffusion due to the assumption of purely magnetic fluctuations, the two-dimensional Fokker–Planck equation can easily be deduced from the more general formula of Schlickeiser (2002, (12.1.26)).

The solution of (1.31), $f = f(\mu, z, t)$, is the ensemble-averaged particle distribution function in the two-dimensional phase-space. It is demonstrated in Shalchi (2005a) that there are early times for which pitch-angle transport has not reached diffusive behavior. The above equation cannot be applied for such early times.

According to the Sturm–Liouville theorem, the solution of the Fokker–Planck equation becomes isotropic for large enough times (pitch-angle isotropization process):

$$f(\mu, z, t \to \infty) \to M(z, t). \tag{1.33}$$

In the following, this property of the solution is applied to derive the diffusion–transport-equation and to calculate the parallel spatial diffusion coefficient.

1.4.2 The Diffusion Equation

By applying the operator $\frac{1}{2}\int_{-1}^{+1} d\mu$ onto the Fokker–Planck equation, we find the continuity equation

$$\frac{\partial M(z,t)}{\partial t} = -\frac{\partial j(z,t)}{\partial z}, \tag{1.34}$$

with the pitch-angle averaged particle density

$$M(z,t) = \frac{1}{2}\int_{-1}^{+1} d\mu \; f(\mu, z, t) \tag{1.35}$$

and the current density

$$j(z,t) = \frac{v}{2}\int_{-1}^{+1} d\mu \; \mu f(\mu, z, t). \tag{1.36}$$

To deduce (1.34), we applied

$$D_{\mu\mu}(\mu = \pm 1) = 0, \tag{1.37}$$

which can directly be obtained from the equations of motion (see e.g., (1.6)). Equation (1.36) can be rewritten as

$$\begin{aligned} j(z,t) &= -\frac{v}{4}\int_{-1}^{+1} d\mu \; \frac{\partial(1-\mu^2)}{\partial \mu} f(\mu, z, t), \\ &\overset{p.I.}{=} \frac{v}{4}\int_{-1}^{+1} d\mu \; (1-\mu^2)\frac{\partial f(\mu, z, t)}{\partial \mu}. \end{aligned} \tag{1.38}$$

To replace the term $\partial f/\partial\mu$, we apply the operator $\int_{-1}^{\mu} d\nu$ onto the Fokker–Planck equation to obtain

$$\frac{\partial}{\partial t}\int_{-1}^{\mu} d\nu\ f(\nu,z,t) + v\frac{\partial}{\partial z}\int_{-1}^{\mu} d\nu\ \nu f(\nu,z,t) = D_{\mu\mu}\frac{\partial f(\mu,z,t)}{\partial\mu}, \tag{1.39}$$

where we used (1.37). By multiplying this equation with $(1-\mu^2)/D_{\mu\mu}$, we find

$$\frac{1-\mu^2}{D_{\mu\mu}}\frac{\partial}{\partial t}\int_{-1}^{\mu} d\nu\ f(\nu,z,t) + \frac{1-\mu^2}{D_{\mu\mu}}v\frac{\partial}{\partial z}\int_{-1}^{\mu} d\nu\ \nu f(\nu,z,t)$$
$$= (1-\mu^2)\frac{\partial f(\mu,z,t)}{\partial\mu}. \tag{1.40}$$

By pitch-angle averaging this result, we find an exact expression for the current density (see (1.38))

$$j(z,t) = \frac{v}{4}\int_{-1}^{+1} d\mu\ \frac{1-\mu^2}{D_{\mu\mu}}\int_{-1}^{\mu} d\nu\ \frac{\partial f(\nu,z,t)}{\partial t}$$
$$+ \frac{v^2}{4}\int_{-1}^{+1} d\mu\ \frac{1-\mu^2}{D_{\mu\mu}}\int_{-1}^{\mu} d\nu\ \nu\frac{\partial f(\nu,z,t)}{\partial z}. \tag{1.41}$$

Now we consider the limit of infinitely late times ($t\to\infty$). Because of the pitch-angle-isotropization process, we have $f(\mu,z,t)\to M(z,t)$ and therefore

$$j(z,t) = \frac{v}{4}\frac{\partial M(z,t)}{\partial t}\int_{-1}^{+1} d\mu\ \frac{1-\mu^2}{D_{\mu\mu}}\int_{-1}^{\mu} d\nu$$
$$+ \frac{v^2}{4}\frac{\partial M(z,t)}{\partial z}\int_{-1}^{+1} d\mu\ \frac{1-\mu^2}{D_{\mu\mu}}\int_{-1}^{\mu} d\nu\ \nu. \tag{1.42}$$

The remaining integrals can easily be solved and we finally find

$$j(z,t) = \frac{v}{4}\frac{\partial M(z,t)}{\partial t}\int_{-1}^{+1} d\mu\ \frac{(1-\mu^2)(1+\mu)}{D_{\mu\mu}}$$
$$- \frac{v^2}{8}\frac{\partial M(z,t)}{\partial z}\int_{-1}^{+1} d\mu\ \frac{(1-\mu^2)^2}{D_{\mu\mu}}. \tag{1.43}$$

With

$$\kappa_{zz} = \frac{v^2}{8}\int_{-1}^{+1} d\mu\ \frac{(1-\mu^2)^2}{D_{\mu\mu}} \tag{1.44}$$

and

$$\kappa_{zt} = \frac{v}{4}\int_{-1}^{+1} d\mu\ \frac{(1-\mu^2)(1+\mu)}{D_{\mu\mu}}, \tag{1.45}$$

we find the expression

$$j(z,t) = \kappa_{zt}\frac{\partial M(z,t)}{\partial t} - \kappa_{zz}\frac{\partial M(z,t)}{\partial z} \tag{1.46}$$

for the current density. To replace the term $\partial M(z,t)/\partial t$, we can use (1.34) to find

$$j(z,t) = -\frac{\partial}{\partial z}\left[\kappa_{zt}\, j(z,t) + \kappa_{zz}M(z,t)\right]. \tag{1.47}$$

For late times we find, by comparing (1.35) with (1.36), that $j(z,t)$ tends to zero for nearly isotropic $f(\mu,z,t)$. Therefore, the current density can be approximated by

$$j(z,t) = -\kappa_{zz}\frac{\partial M(z,t)}{\partial z}. \tag{1.48}$$

Combining this result with the continuity equation (1.34), we obtain

$$\frac{\partial M(z,t)}{\partial t} = \kappa_{zz}\frac{\partial^2 M(z,t)}{\partial z^2} \tag{1.49}$$

to lowest order of the pitch-angle anisotropy $f(\mu,z,t) - M(z,t)$. Equation (1.49) is a diffusion equation for the pitch-angle averaged distribution function $M(z,t)$.

Now it is straightforward to calculate the moments of $M(z,t)$. By multiplying the diffusion equation with z and by integrating the result, we find

$$\frac{\mathrm{d}}{\mathrm{d}t}\langle z\rangle = \kappa_{zz}\int_{-\infty}^{\infty}\mathrm{d}z\, z\frac{\partial^2 M(z,t)}{\partial z^2} \overset{p.I.}{=} -\kappa_{zz}\int_{-\infty}^{\infty}\mathrm{d}z\,\frac{\partial M(z,t)}{\partial z} = 0 \tag{1.50}$$

with

$$\langle z\rangle(t) = \int_{-\infty}^{\infty}\mathrm{d}z\, zM(z,t), \tag{1.51}$$

where we assumed $z(\partial^2 M/\partial z^2)(z \to \pm\infty) \to 0$ and $M(z \to \pm\infty) \to 0$. Obviously, we find $\langle z\rangle(t \to \infty) \to$ const, which is the expected result due to the pitch-angle isotropization process. Similarly, we can derive:

$$\frac{\mathrm{d}}{\mathrm{d}t}\left\langle z^2\right\rangle = \kappa_{zz}\int_{-\infty}^{\infty}\mathrm{d}z\, z^2\frac{\partial^2 M(z,t)}{\partial z^2} \overset{p.I.}{=} -2\kappa_{zz}\int_{-\infty}^{\infty}\mathrm{d}z\, z\frac{\partial M(z,t)}{\partial z}$$
$$\overset{p.I.}{=} 2\kappa_{zz}\int_{-\infty}^{\infty}\mathrm{d}z\, M(z,t) = 2\kappa_{zz}, \tag{1.52}$$

where we assumed $z^2(\partial^2 M/\partial z^2)(z \to \pm\infty) \to 0$ and $z(\partial M/\partial z)(z \to \pm\infty) \to 0$. Furthermore, we applied the normalization constraint

$$\int_{-\infty}^{\infty}\mathrm{d}z\, M(z,t) = 1. \tag{1.53}$$

Obviously, the running diffusion coefficient (defined by (1.21)) is

$$d_{zz}(t) = \frac{1}{2}\frac{\mathrm{d}\langle z^2\rangle}{\mathrm{d}t} = \kappa_{zz} = \text{const} \tag{1.54}$$

and the parameter $\kappa_{zz} = \kappa_{\parallel}$ can be identified with the parallel spatial diffusion coefficient. In (1.44), a formula for the parallel spatial diffusion coefficient has been derived, which relates the pitch-angle Fokker–Planck coefficient $D_{\mu\mu}$ to the parallel spatial diffusion coefficient $\kappa_{\parallel}$ or to the parallel mean free path $\lambda_{\parallel}$

$$\lambda_{\parallel} = \frac{3}{v}\kappa_{\parallel} = \frac{3v}{8}\int_{-1}^{+1} \mathrm{d}\mu \, \frac{(1-\mu^2)^2}{D_{\mu\mu}}. \tag{1.55}$$

Different derivations of this relation have been presented by Jokipii (1966), Hasselmann and Wibberenz (1968), Earl (1974), and Shalchi (2006a) in a different way.

1.4.3 Solution of the Diffusion Equation

Here we solve the diffusion equation (see (1.49)) derived in the previous paragraph. First, we apply the *ansatz* $M(z,t) = \rho(t)P(z)$ to find

$$\frac{1}{\kappa_{\parallel}\rho(t)}\frac{\partial \rho(t)}{\partial t} = \frac{1}{P(z)}\frac{\partial^2 P(z)}{\partial z^2} =: -k^2, \tag{1.56}$$

where we introduced the (real) constant parameter k. From (1.56), we can deduce two ordinary differential equations, which can easily be solved

$$\begin{aligned} \frac{\partial \rho(t)}{\partial t} &= -k^2\kappa_{\parallel}\rho(t) \Rightarrow \rho(t) \sim \mathrm{e}^{-k^2\kappa_{\parallel}t}, \\ \frac{\partial^2 P(z)}{\partial z^2} &= -k^2 P(z) \Rightarrow P(z) \sim \mathrm{e}^{\mathrm{i}kz}, \end{aligned} \tag{1.57}$$

and therefore the general solution of the diffusion equation is

$$M(z,t) = \int \mathrm{d}k \; a(k)\,\mathrm{e}^{-k^2\kappa_{\parallel}t+\mathrm{i}kz}. \tag{1.58}$$

The function $a(k)$ can easily be expressed by the initial distribution function $M(z,0)$. Using

$$M(z,0) = \int \mathrm{d}k \; a(k)\,\mathrm{e}^{\mathrm{i}kz} \tag{1.59}$$

we can deduce

$$\int dz\, M(z,0)\, e^{-ik' z} = \int dk\, a(k) \int dz\, e^{i(k-k')z} = 2\pi a(k'). \tag{1.60}$$

This relation can be combined with (1.58) to get

$$M(z,t) = \frac{1}{2\pi} \int dk \int dz'\; M(z',0)\, e^{-k^2 \kappa_\parallel t + ik(z-z')}. \tag{1.61}$$

The integral over k can be solved

$$\int dk\, e^{-k^2 \kappa_\parallel t + ik(z-z')} = \sqrt{\frac{\pi}{\kappa_\parallel t}}\, e^{-\frac{(z-z')^2}{4\kappa_\parallel t}} \tag{1.62}$$

to find the form

$$M(z,t) = \frac{1}{\sqrt{4\pi\kappa_\parallel t}} \int dz'\; M(z',0)\, e^{-\frac{(z-z')^2}{4\kappa_\parallel t}}. \tag{1.63}$$

Equation (1.63) is the general solution of the diffusion equation. As an example, we consider a sharp initial distribution function of the particle

$$M(z',0) = \delta(z'). \tag{1.64}$$

In this case, (1.63) can easily be evaluated and we find a Gaussian distribution

$$M_G(z,t) = \frac{1}{\sqrt{4\pi\kappa_\parallel t}}\, e^{-\frac{z^2}{4\kappa_\parallel t}}, \tag{1.65}$$

which is the solution of the diffusion equation for sharp initial conditions. Equation (1.65) is used in most of the applications discussed in this book (see Chaps. 4–7).

1.5 The Physics of Perpendicular Scattering

Here, we discuss the physics of particle transport perpendicular to the mean field $\mathbf{B}_0$. The starting point here are the perpendicular components of the equations of motion (see (1.6)), which can be written as

$$\begin{aligned} \dot{v}_x &= \Omega v_y + b_y, \\ \dot{v}_y &= -\Omega v_x + b_x, \end{aligned} \tag{1.66}$$

with the perturbation terms

$$b_x = \Omega\left(v_z\frac{\delta B_x}{B_0} - v_x\frac{\delta B_z}{B_0}\right),$$

$$b_y = \Omega\left(v_y\frac{\delta B_z}{B_0} - v_z\frac{\delta B_y}{B_0}\right). \tag{1.67}$$

It is a simple matter to demonstrate that the homogeneous solution of (1.66) can be written as

$$v_x^h(t) = A\,\cos(\Omega t) + B\,\sin(\Omega t),$$

$$v_y^h(t) = B\,\cos(\Omega t) - A\,\sin(\Omega t). \tag{1.68}$$

To find a particular solution of (1.66), we substitute the parameters A and B in (1.68) by $A \to A(t)$ and $B \to B(t)$ to obtain

$$v_x(t) = A(t)\,\cos(\Omega t) + B(t)\,\sin(\Omega t),$$

$$v_y(t) = B(t)\,\cos(\Omega t) - A(t)\,\sin(\Omega t). \tag{1.69}$$

With this *ansatz*, (1.66) becomes

$$\dot{A}\,\cos(\Omega t) + \dot{B}\,\sin(\Omega t) = b_y,$$

$$\dot{B}\,\cos(\Omega t) - \dot{A}\,\sin(\Omega t) = b_x. \tag{1.70}$$

This set of equations can easily be solved and we find

$$A(t) = \int_0^t \mathrm{d}\tau\,\left[b_y(\tau)\,\cos(\Omega\tau) - b_x(\tau)\,\sin(\Omega\tau)\right],$$

$$B(t) = \int_0^t \mathrm{d}\tau\,\left[b_y(\tau)\,\sin(\Omega\tau) + b_x(\tau)\,\cos(\Omega\tau)\right]. \tag{1.71}$$

Then the general solution of (1.66) for the x component is

$$\begin{aligned} v_x(t) = {} & v_x(0)\,\cos(\Omega t) + v_y(0)\,\sin(\Omega t) \\ & + \cos(\Omega t)\int_0^t \mathrm{d}\tau\,\left[b_y(\tau)\,\cos(\Omega\tau) - b_x(\tau)\,\sin(\Omega\tau)\right] \\ & + \sin(\Omega t)\int_0^t \mathrm{d}\tau\,\left[b_y(\tau)\,\sin(\Omega\tau) + b_x(\tau)\,\cos(\Omega\tau)\right] \end{aligned} \tag{1.72}$$

and for the y component

$$
\begin{aligned}
v_y(t) = {} & v_y(0)\,\cos(\Omega t) - v_x(0)\,\sin(\Omega t) \\
& + \cos(\Omega t) \int_0^t \mathrm{d}\tau\ \left[b_y(\tau)\,\sin(\Omega\tau) + b_x(\tau)\,\cos(\Omega\tau)\right] \\
& - \sin(\Omega t) \int_0^t \mathrm{d}\tau\ \left[b_y(\tau)\,\cos(\Omega\tau) - b_x(\tau)\,\sin(\Omega\tau)\right]. \qquad (1.73)
\end{aligned}
$$

To proceed, we calculate the guiding center velocity, which can be defined as

$$
\tilde{v}_i(\xi) := \frac{1}{T} \int_\xi^{\xi+T} \mathrm{d}t\ v_i(t), \qquad (1.74)
$$

where we used

$$
T = \frac{2\pi}{\Omega}. \qquad (1.75)
$$

It can easily be demonstrated that

$$
\begin{aligned}
\tilde{v}_x(\xi) = {} & \frac{1}{2\pi} \int_\xi^{\xi+T} \mathrm{d}t\ \left[\frac{\mathrm{d}}{\mathrm{d}t}\ \sin(\Omega t)\right] \\
& \times \int_0^t \mathrm{d}\tau\ \left[b_y(\tau)\,\cos(\Omega\tau) - b_x(\tau)\,\sin(\Omega\tau)\right] \\
& - \frac{1}{2\pi} \int_\xi^{\xi+T} \mathrm{d}t\ \left[\frac{\mathrm{d}}{\mathrm{d}t}\ \cos(\Omega t)\right] \\
& \times \int_0^t \mathrm{d}\tau\ \left[b_y(\tau)\,\sin(\Omega\tau) + b_x(\tau)\,\cos(\Omega\tau)\right]. \qquad (1.76)
\end{aligned}
$$

By using the method of partial integration, we obtain

$$
\begin{aligned}
\tilde{v}_x(t) = {} & \frac{1}{2\pi} \int_t^{t+T} \mathrm{d}\tau\ b_x(\tau) \\
& + \frac{\sin(\Omega t)}{2\pi} \int_t^{t+T} \mathrm{d}\tau\ \left[b_y(\tau)\,\cos(\Omega\tau) - b_x(\tau)\,\sin(\Omega\tau)\right] \\
& - \frac{\cos(\Omega t)}{2\pi} \int_t^{t+T} \mathrm{d}\tau\ \left[b_y(\tau)\,\sin(\Omega\tau) + b_x(\tau)\,\cos(\Omega\tau)\right]. \qquad (1.77)
\end{aligned}
$$

Analogously, we can derive the y component for the guiding center velocity

$$
\tilde{v}_y(t) = \frac{-1}{2\pi} \int_t^{t+T} \mathrm{d}\tau\ b_y(\tau)
$$

$$+ \frac{\sin(\Omega t)}{2\pi} \int_t^{t+T} d\tau \, \left[b_y(\tau) \, \sin(\Omega \tau) + b_x(\tau) \, \cos(\Omega \tau) \right]$$

$$+ \frac{\cos(\Omega t)}{2\pi} \int_t^{t+T} d\tau \, \left[b_y(\tau) \, \cos(\Omega \tau) - b_x(\tau) \, \sin(\Omega \tau) \right]. \quad (1.78)$$

To proceed, we consider the limit of slowly varying perturbation terms:

$$b_i(t) \approx b_i(t + \xi) \quad \text{for} \quad 0 \leq \xi \leq T. \quad (1.79)$$

In this case we find

$$\tilde{v}_x(t) \approx \frac{1}{\Omega} b_x(t) = v_z \frac{\delta B_x}{B_0} - v_x \frac{\delta B_z}{B_0},$$

$$\tilde{v}_y(t) \approx -\frac{1}{\Omega} b_y(t) = v_z \frac{\delta B_y}{B_0} - v_y \frac{\delta B_z}{B_0}. \quad (1.80)$$

Within the turbulence models considered in this book (slab-, pure 2D-, composite-model, see Chap. 2), we have $\delta B_z = 0$, and hence we find

$$\tilde{v}_x(t) \approx v_z \frac{\delta B_x}{B_0},$$

$$\tilde{v}_y(t) \approx v_z \frac{\delta B_y}{B_0}. \quad (1.81)$$

By assuming that

$$\tilde{v}_z(t) \approx v_z(t), \quad (1.82)$$

we find that guiding centers follow magnetic field lines (see Fig. 1.4). Therefore, the particle motion can be considered as a superposition of a gyrorotation of the particle and a stochastic motion of the guiding center, which follows magnetic field lines. When calculating a diffusion coefficient, the gyromotion can be neglected; thus, the particle diffusion coefficient can be approximated by the guiding center diffusion coefficient if (1.79) holds. To determine the perpendicular diffusion coefficient, (1.81) can be combined with the TGK formulation (1.29):

$$\kappa_{ij} = \int_0^\infty dt \, \left\langle \tilde{v}_i(t) \tilde{v}_j(0) \right\rangle,$$

$$= \frac{1}{B_0^2} \int_0^\infty dt \, \langle \tilde{v}_z(t) \tilde{v}_z(0) \delta B_x(\mathbf{x}(t), t) \delta B_x(\mathbf{x}(0), 0) \rangle, \quad (1.83)$$

with $i, j = x, y$.

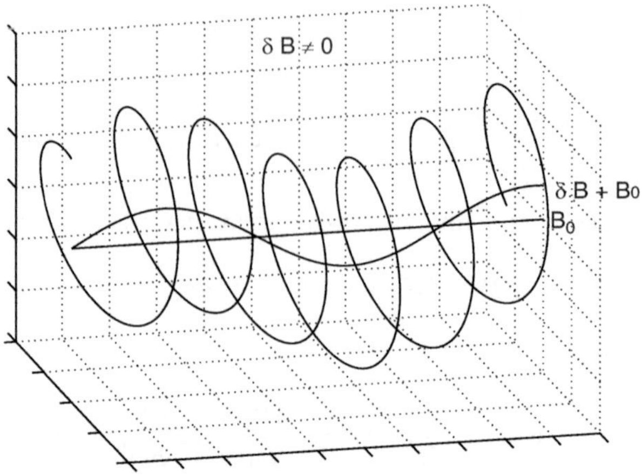

Fig. 1.4 If the guiding centers are following magnetic field lines, the particle experiences perpendicular scattering due to the turbulent character of the field lines

1.6 The Diffusion Tensor and Momentum Diffusion

The transport of cosmic rays through a magnetized astrophysical plasma can be described by the diffusion tensor

$$\left(\kappa_{ij}\right) = \begin{pmatrix} \kappa_\perp & \kappa_{\mathrm{AS}} & 0 \\ -\kappa_{\mathrm{AS}} & \kappa_\perp & 0 \\ 0 & 0 & \kappa_\parallel \end{pmatrix}. \tag{1.84}$$

The diagonal elements describe parallel (see Sect. 1.4) as well as perpendicular diffusion (Sect. 1.5). The off-diagonal, antisymmetric terms (κ_{AS}) describe effects of gradient and curvature drifts (see Jokipii et al. 1977; Bieber and Matthaeus 1997).

Although one could suspect that some interesting effects may have their origin in turbulence models that lead to $\kappa_{\mathrm{AS}} \neq 0$, we shall here apply various transport theories only to systems with $\kappa_{\mathrm{AS}} = 0$. In that case, we have

$$\left(\kappa_{ij}\right) = \begin{pmatrix} \kappa_\perp & 0 & 0 \\ 0 & \kappa_\perp & 0 \\ 0 & 0 & \kappa_\parallel \end{pmatrix} \tag{1.85}$$

so that only knowledge of two parameters is necessary to describe spatial diffusion.

In Sect. 1.4, a one-dimensional diffusion equation has been derived by neglecting perpendicular diffusion. By investigating the full six-dimensional Fokker–Planck equation, a three-dimensional diffusion equation can be derived (Schlickeiser 2002, (12.3.21)). For axisymmetric turbulence ($\kappa_{xy} = \kappa_{yx} = 0$) and for purely magnetic fluctuations (no momentum diffusion), we find a diffusion equation of the form

$$\frac{\partial M(\mathbf{x},t)}{\partial t} = \sum_{i=x,y,z} \kappa_{ii} \frac{\partial^2 M(\mathbf{x},t)}{\partial x_i^2}. \tag{1.86}$$

It is a simple matter to demonstrate that the general solution of (1.86) is (see e.g., (1.63))

$$M(\mathbf{x},t) = \frac{1}{(2\pi)^{3/2}\sqrt{8t^3\kappa_{xx}\kappa_{yy}\kappa_{zz}}} \int \mathrm{d}^3x' \; M(\mathbf{x}',0) \prod_{i=x,y,z} \mathrm{e}^{-\frac{(x_i - x_i')^2}{4t\kappa_{ii}}}. \tag{1.87}$$

For sharp initial conditions ($M(\mathbf{x}',0) = \delta(\mathbf{x}')$), we find a Gaussian function of the form

$$M_G(\mathbf{x},t) = \frac{1}{(2\pi)^{3/2}\sqrt{8t^3\kappa_{xx}\kappa_{yy}\kappa_{zz}}} \mathrm{e}^{-\frac{x^2}{4t\kappa_{xx}} - \frac{y^2}{4t\kappa_{yy}} - \frac{z^2}{4t\kappa_{zz}}}. \tag{1.88}$$

The moments of the distribution function can easily be deduced:

$$\left\langle x_i^2 \right\rangle = 2t\kappa_{ii}. \tag{1.89}$$

For nondiffusive processes such as subdiffusive perpendicular transport, the diffusion equation is no longer valid. In this case, the diffusion equation has to be replaced by another differential equation, for example, by the Telegrapher's equation. Furthermore, the Fokker–Planck equation has to be substituted by a fractional Fokker–Planck equation. Some aspects of such nondiffusive processes and consequences for the different transport equations are discussed in Webb et al. (2006), where a Chapman–Kolmogorov equation description of cosmic ray transport has been applied.

For the formulation of a nonlinear particle transport theory, such as the nonlinear guiding center theory (Matthaeus et al. 2003, see Chap. 4), the characteristic function

$$\Gamma(\mathbf{k},t) = \left\langle \mathrm{e}^{\mathrm{i}\mathbf{k}\cdot\mathbf{x}} \right\rangle \tag{1.90}$$

is a key input. For the Gaussian particle distribution of (1.88), we have

$$\begin{aligned} \Gamma(\mathbf{k},t) &= \frac{1}{(2\pi)^3} \int \mathrm{d}^3k \; \mathrm{e}^{-\sum_{i=x,y,z} k_i^2 \kappa_{ii} t} \int \mathrm{d}^3x \; \mathrm{e}^{i(\mathbf{k}' - \mathbf{k})\mathbf{x}}, \\ &= \int \mathrm{d}^3k \; \mathrm{e}^{-\sum_{i=x,y,z} k_i^2 \kappa_{ii} t} \delta(\mathbf{k}' - \mathbf{k}), \end{aligned} \tag{1.91}$$

and we finally obtain

$$\Gamma(\mathbf{k},t) = \mathrm{e}^{-\kappa_{xx} k_x^2 t - \kappa_{yy} k_y^2 t - \kappa_{zz} k_z^2 t}. \tag{1.92}$$

For axisymmetric turbulence, we have $\kappa_\perp = \kappa_{xx} = \kappa_{yy}$ and, thus,

$$\Gamma(\mathbf{k}, t) = \mathrm{e}^{-\kappa_\perp k_\perp^2 t - \kappa_\| k_\|^2 t}. \tag{1.93}$$

By assuming that the Gaussian distribution is a good approximation for the real particle distribution function, (1.93) can be used to approximate the characteristic function. In Chaps. 4–7, this idea will be used to formulate nonlinear transport theories.

1.6.1 Fokker–Planck vs. Diffusion Coefficients

In the definitions of spatial diffusion coefficients, the ensemble averaging operator $\langle \ldots \rangle$ has been used. In (1.83), this operator is a total average, which means that it is also averaged over the pitch-angle. Alternatively, one can describe particle transport in a six-dimensional phase-space, where the diffusion coefficients are pitch-angle dependent (μ-dependent) and are usually called the Fokker–Planck coefficients:

$$\begin{aligned} D_{\mu\mu}(\mu) &= \lim_{t\to\infty} \frac{\langle (\Delta\mu)^2 \rangle}{2t} = \int_0^\infty \mathrm{d}t \ \langle \dot{\mu}(t)\dot{\mu}(0) \rangle, \\ D_{ij}(\mu) &= \lim_{t\to\infty} \frac{\langle (\Delta x_i)^2 \rangle}{2t} = \int_0^\infty \mathrm{d}t \ \langle \tilde{v}_i(t)\tilde{v}_j(0) \rangle, \quad i, j = x, y. \end{aligned} \tag{1.94}$$

The main reason for describing pitch-angle dependent particle scattering is the relation (1.55). To understand transport along the mean magnetic field, one has to compute the pitch-angle-dependent Fokker–Planck coefficient $D_{\mu\mu}(\mu)$. For describing scattering across the mean field, it is often much easier to calculate the (pitch-angle averaged) spatial diffusion coefficients directly (see e.g., Chap. 4).

The phase-space particle distribution function can be obtained by solving the Fokker–Planck equation, which is the transport equation in the six-dimensional space. The pitch-angle Fokker–Planck coefficient $D_{\mu\mu}$ controls the parallel spatial diffusion coefficient via (1.55); the perpendicular spatial diffusion coefficients are

$$\kappa_{ij} = \frac{1}{2} \int_{-1}^{+1} \mathrm{d}\mu \ D_{ij}(\mu), \quad i, j = x, y. \tag{1.95}$$

This formulation is applicable in a quasilinear description of cosmic ray transport as discussed in Chap. 3. Some parts of this formulation such as (1.55) will be applied also in nonlinear transport theories (see Chaps. 4–7).

1.6.2 Cosmic Ray Momentum Diffusion Due to Electric Fields

In this book, only turbulent systems with purely magnetic fluctuations are discussed and electric fields are neglected ($\delta E_i = 0$). In reality, however, these fields do exist. Although they are assumed to be less important for spatial diffusion, they are essential for momentum diffusion (stochastic acceleration due to the turbulence). By multiplying (1.1) with the momentum vector $\mathbf{p}$, it can easily be demonstrated that

$$\mathbf{p} \cdot \frac{\mathrm{d}\mathbf{p}}{\mathrm{d}t} = q\mathbf{p} \cdot \mathbf{E} + \mathbf{p} \cdot \left(\frac{\mathbf{v}}{c} \times \mathbf{B} \right) = q\mathbf{p} \cdot \delta\mathbf{E}. \tag{1.96}$$

The left hand side can be written as $p\dot{p}$ and therefore

$$\dot{p} = \frac{q}{p}\mathbf{p} \cdot \delta\mathbf{E}, \tag{1.97}$$

and in terms of the velocity

$$\dot{p} = \frac{q}{v}\mathbf{v} \cdot \delta\mathbf{E} = \frac{q}{v}\left(v_x \delta E_x + v_y \delta E_y + v_z \delta E_z \right). \tag{1.98}$$

As demonstrated in Schlickeiser (2002, (12.3.24)–(12.3.30)), the momentum diffusion coefficient A can be obtained by applying

$$A = \frac{1}{2} \int_{-1}^{+1} \mathrm{d}\mu \left(D_{pp} - \frac{D_{\mu p}^2}{D_{\mu\mu}} \right). \tag{1.99}$$

The Fokker–Planck coefficients $D_{\mu\mu}$, $D_{\mu p}$, and D_{pp} can be calculated by combining (1.98) with the TGK formulation. Although not done here, it is straightforward to calculate momentum diffusion coefficients by applying a transport theory such as quasilinear theory (see Chap. 3). In this case, however, also the correlation tensor of the electric fields has to be specified (see Chap. 2). A detailed review of quasilinear momentum diffusion in various turbulence models (e.g., shear Alfvénic or fast magnetosonic turbulence) can be found in Schlickeiser (2002).

1.7 Cosmic Ray Mean Free Paths Deduced from Observations

The main goal of cosmic ray transport theory is the explanation of observations. In this section, some prominent examples for measured mean free paths are discussed. First, cosmic ray transport in the heliosphere is considered, where mean free paths can be extracted from observations. As a second example, we discuss particle transport in the interstellar medium, where the parallel mean free path can be obtained from measurements of the primary-to-secondary abundance ratio.

1.7.1 Observed Mean Free Paths in the Heliosphere

In the solar wind, time-intensity and time-anisotropy profiles of charged particles can be measured. From these observations, cosmic ray mean free paths can be deduced by applying certain transport models and assumptions. In most cases (sufficiently strong scattering), the propagation can be modeled by a process of spatial diffusion. Thus, a diffusion equation can be combined with the measured profiles to extract scattering mean free paths.

In a famous and influential review article, Palmer (1982) compared in detail different mean free paths derived from electron and proton observations. By assembling observational results from Lin (1970, 1974), Palmer et al. (1975, 1978), McCarthy and O'Gallagher (1976), Ford et al. (1977), Hamilton (1977), Schulze et al. (1977), Zwickl and Webber (1977, 1978), Ma Sung and Earl (1978), Bieber et al. (1980), Chenette (1980), Debrunner and Lockwood (1980), Potter et al. (1980), and Lockwood et al. (1982), the results for the parallel mean free path shown in Fig. 1.5 can be obtained. Here, the cosmic ray parallel mean free paths are shown in astronomical units[2] (AU) as a function of the particle rigidity, which is defined as

$$\text{Rigidity} = \frac{pc}{\mid q \mid} \tag{1.100}$$

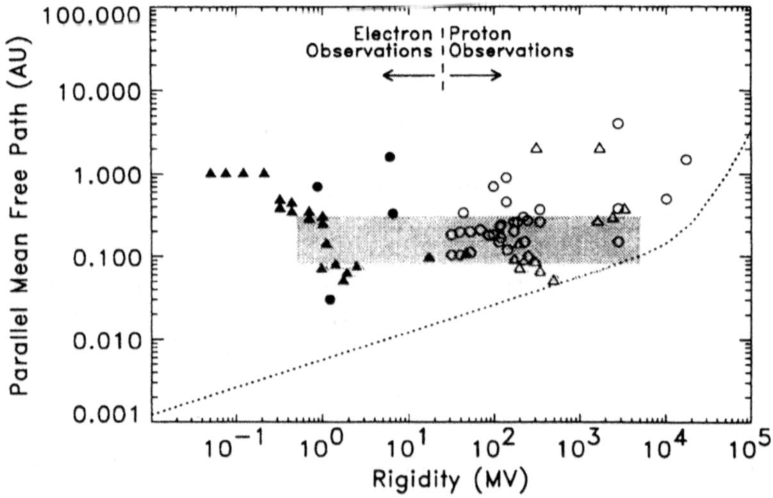

Fig. 1.5 Parallel mean free paths as a function of the cosmic ray rigidity. The different symbols denote results derived from electron and proton observations. The *shaded band* is the observational consensus enunciated by Palmer (1982). The *dotted line* represents the prediction of standard quasi-linear theory (see Chap. 3 for a detailed discussion of this approach). Reprinted with permission from The American Astronomical Society, Bieber et al. (1994)

[2] An astronomical unit is approximately equal to the average distance between the Earth and the Sun. 1 AU $\approx 149.60 \times 10^9$ m.

in megavolt (MV) units. In (1.100), the speed of light, c, the momentum, p, and the charge, q, have been used. Palmer suggested to approximate these measurement by the box also shown in Fig. 1.5. This box, often referred to as the *Palmer consensus range*, reaches from quite small rigidities (0.5 MV) up to large values where the particle gyroradius becomes comparable to the correlation length of the turbulence (5.000 MV). In this rigidity range, the parallel mean free path can reach quite different values. Most observations indicate

$$0.08\,\mathrm{AU} \leq \lambda_{\parallel} \leq 0.3\,\mathrm{AU}. \tag{1.101}$$

From Ulysses observations, Gloeckler et al. (1995) concluded that the parallel mean free path for pickup ions is 2 AU at 2.4 MV rigidity (they stated conservatively that $\lambda_{\parallel}$ is of order 1 AU, but actually they obtained the best fit for 2 AU). It should be noted, however, that this observation was taken at high heliographic latitudes and at a heliocentric distance of 2.34 AU, whereas the Palmer box indicates the parallel mean free path at 1 AU; these differences should be kept in mind when comparing with other observations at 1 AU heliocentric distance.

Möbius et al. (1998) concluded from AMPTE (Active Magnetospheric Particle Tracer Explorers) spacecraft observations that the parallel mean free path of pickup helium ranges from 0.16 to 0.76 AU at 5.6 MV rigidity in the data they analyzed. In Fig. 1.6, the Möbius et al. (1998) results are shown together with the Gloeckler et al. (1995) observations and the Palmer consensus range.

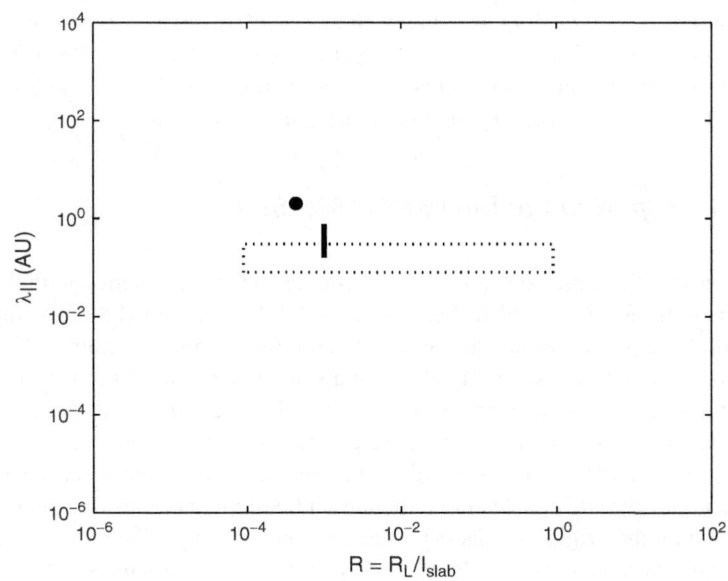

Fig. 1.6 The parallel mean free path $\lambda_{\parallel}$ from observations: Palmer consensus range (Palmer 1982, *box*), Ulysses observations (Gloeckler et al. 1995, *dot*), and AMPTE spacecraft observations (Möbius et al. 1998, *vertical line*)

Also the perpendicular mean free path can be obtained from different observations. In applications of cosmic ray diffusion theory such as modulation modeling, the perpendicular diffusion coefficient is often taken to be a fixed ratio of the parallel and the perpendicular coefficient:

$$\frac{\lambda_\perp}{\lambda_\parallel} = b, \tag{1.102}$$

with b a constant typically taken to be in the range $0.005 \leq b \leq 0.05$ (e.g., Jokipii et al. 1995, Ferrando 1997, Burger et al. 2000, Ferreira et al. 2001). Such values can be obtained from certain observations. For instance, Palmer (1982) suggested that the perpendicular mean free path at 1 AU heliocentric distance is $\lambda_\perp \approx 0.0067\,\text{AU}$. However, Palmer pointed out that the spreading around this average is rather large. Together with (1.101), we find approximately

$$0.02 \leq \frac{\lambda_\perp}{\lambda_\parallel} \leq 0.08. \tag{1.103}$$

Chenette et al. (1977) deduced perpendicular mean free paths from Jovian electrons and they concluded that $\lambda_\perp \approx 0.005\,\text{AU}$. More recently, Burger et al. (2000) derived from Ulysses measurement of galactic protons that $\lambda_\perp \approx 0.008\,\text{AU}$ at higher rigidities. Other authors (Dwyer et al. 1997, Zhang et al. 2003) noted that $\lambda_\perp/\lambda_\parallel$ is occasionally very large, even approaching or exceeding unity. In Fig. 1.7, the different observations for perpendicular mean free paths are shown and can easily be compared with each other. One of the goals of cosmic ray transport theories is to reproduce the observed mean free paths visualized in Figs. 1.5–1.7. It will be demonstrated in Chap. 8 that advanced models for the turbulence and nonlinear transport theories are necessary to explain these measurements.

1.7.2 Transport in the Interstellar Medium

The properties of cosmic rays, their composition, and their interaction with the interstellar medium are discussed in Schlickeiser (2002). Some of the most important aspects of these properties are summarized in the following paragraph.

Cosmic rays include essentially all elements of the periodic table. Approximately 89% of the nuclei are hydrogen (protons), 10% helium, and about 1% heavier elements. The common heavier elements (e.g., carbon, oxygen, magnesium, silicon, and iron) are present in the same relative abundances as in the solar system, but there are important differences in elemental and isotopic composition that provide information on the origin and history of galactic cosmic rays. For example, there is a significant overabundance of the elements Li, Be, and B produced when heavier cosmic rays such as carbon, nitrogen, and oxygen fragment into lighter nuclei during collisions with the interstellar gas. Similar collisions of cosmic ray nuclei produce lighter nuclear fragments, including radioactive isotopes such as ^{10}Be, which has a

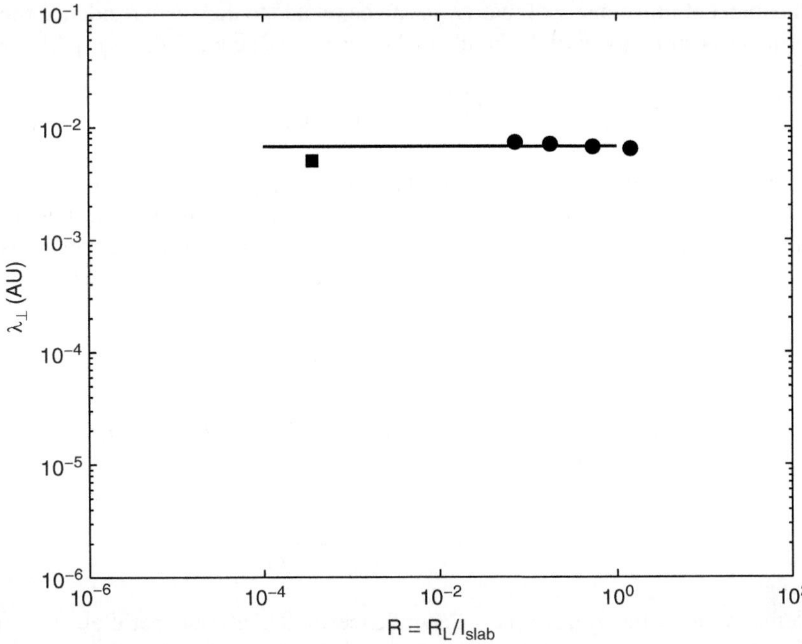

Fig. 1.7 The perpendicular mean free path $\lambda_\perp$ from observations. Shown are Jovian electrons (Chenette et al. 1977, *square*), Ulysses measurements of galactic protons (Burger et al. 2000, *dots*), and the Palmer (1982, *horizontal line*) value

half-life of 1.6 million years. The measured amount of ^{10}Be in cosmic rays implies that, on average, cosmic rays spend about 10^7 years in the Galaxy before escaping into the intergalactic space.

The number of the different particles can be measured as a function of energy (see e.g., Chap. 3.4 of Schlickeiser 2002). From these results, the relative element abundances can be obtained. The solar system abundance is representative of a sample matter produced in the interiors of stars, and these elements are also seen in optical lines in stellar spectra. It can be recognized by comparing these measurements with the solar system abundance that the cosmic ray abundances are very different from the solar system abundances.

The reason for these strong abundance variations in the arriving cosmic rays finds a natural explanation in the spallation of heavier nuclei during their propagation from their source to the solar system. Thus, the observed abundance can be used to determine the amount of matter through which the cosmic ray particles have had to pass during their journey to the solar system. The cosmic particles, which are produced in stellar sources, are referred to as primary cosmic rays, whereas the measured elements are called secondary cosmic rays.

The total column density of matter, ρ, can be related to the mean residence time of the primary cosmic ray particles in the Galaxy τ via (see e.g., (3.4.1) of Schlickeiser 2002).

$$\rho = \int_0^\infty \mathrm{d}x \; n(x) \approx n_0 v \tau, \tag{1.104}$$

where v is the cosmic ray velocity and n_0 is the average density of the interstellar gas. To proceed, we assume that the particle propagation through the interstellar medium behaves diffusively. Furthermore, we assume that parallel scattering is the dominant process and controls the mean residence time. Thus, we can use

$$\kappa_\parallel = \frac{\left\langle (\Delta z)^2 \right\rangle}{2t} \approx \frac{L^2}{2\tau} \sim \tau^{-1} \tag{1.105}$$

with the system size (thickness of the galactic disk) L. With (1.104) and the relation $\lambda_\parallel = 3\kappa_\parallel / v$, we can deduce

$$\rho(R) \sim v\tau \sim \frac{v}{\kappa_\parallel} \sim \frac{1}{\lambda_\parallel(R)}. \tag{1.106}$$

As demonstrated, the rigidity (R) dependence of the inverse parallel mean free path controls the rigidity dependence of the total column density of matter, $\rho(R)$. By measuring the decrease of the abundance ratio of primary-to-secondary cosmic ray nuclei as B/C and N/O at kinetic energies above 1 GeV/nucleon, Swordy et al. (1990) stated that the variation of the total column density as a function of rigidity R must be

$$\rho(R) = 6.9\,(R/[20\ \mathrm{GV/Nucleon}])^{-a}\ \mathrm{g\,cm}^{-2},\ a = 0.6 \pm 0.1. \tag{1.107}$$

Because the rigidity dependence and, therefore, the parameter a is controlled by the rigidity dependence of the inverse parallel mean free path (see (1.106)), we can determine the rigidity dependence of the parallel mean free path of galactic cosmic rays:

$$\lambda_\parallel(R) \sim R^a,\ a = 0.6 \pm 0.1. \tag{1.108}$$

In Chap. 3, it will be demonstrated that the standard approach (application of quasi-linear theory) predicts $a = 1/3$, which is not in accord with (1.108). However, in Chap. 3 it will also be demonstrated by using test-particle simulations that, for realistic turbulence models, a linear description of parallel diffusion is not appropriate. Therefore, a nonlinear theory for parallel diffusion has to be developed. In Chap. 5, such an improved theory, namely the weakly nonlinear theory, will be derived, and in Chap. 8, this theory will be applied onto transport of galactic cosmic rays to demonstrate that we can indeed reproduce (1.108).

Chapter 2
On Astrophysical Turbulence

Particle scattering in turbulent magnetic systems such as the solar wind or the interstellar medium depends on the properties of the turbulent fields δB_i. In this chapter, it is demonstrated that the turbulence geometry, the wave spectrum, and the temporal behavior of these fields have to be specified. For the geometry, we employ the so-called slab/2D composite model, which is often used to approximate solar wind turbulence. Correlation functions for slab, pure 2D, and composite turbulence models are calculated, as well as field line diffusion coefficients. Although for slab geometry, the field line random walk can be described linearly, for more realistic models, a nonlinear formulation has to be applied. In this chapter, different approaches to describe field line random walk linearly and nonlinearly are discussed. These results are the basis to describe theoretically particle transport in partially turbulent magnetic systems.

2.1 General Forms of the Magnetic Correlation Tensor

The key input into cosmic ray transport theories is the (two-point-two-time) correlation tensor, whose components are

$$R_{lm}(\mathbf{x}, t, \mathbf{x}_0, t_0) = \langle \delta B_l(\mathbf{x}, t) \delta B_m^*(\mathbf{x}_0, t_0) \rangle. \tag{2.1}$$

For applications in transport theory, it is convenient to introduce the correlation tensor in the wavenumber space. By using the Fourier representation

$$\delta B_l(\mathbf{x}, t) = \int d^3k \; \delta B_l(\mathbf{k}, t) \, e^{i\mathbf{k}\cdot\mathbf{x}}, \tag{2.2}$$

we find

$$R_{lm}(\mathbf{x}, t, \mathbf{x}_0, t_0) = \int d^3k \int d^3k' \; \left\langle \delta B_l(\mathbf{k}, t) \delta B_m^*(\mathbf{k}', t_0) \right\rangle \, e^{i\mathbf{k}\cdot\mathbf{x} - i\mathbf{k}'\cdot\mathbf{x}_0}. \tag{2.3}$$

A. Shalchi, *Nonlinear Cosmic Ray Diffusion Theories*,
Astrophysics and Space Science Library,
DOI: 10.1007/978-3-642-00309-7_2,

In this book, only homogeneous turbulence is considered, where

$$\left\langle \delta B_l(\mathbf{k},t)\delta B_m(\mathbf{k}',t_0)\right\rangle = P_{lm}(\mathbf{k},t,t_0)\delta(\mathbf{k}-\mathbf{k}'), \tag{2.4}$$

with the correlation tensor in the wavenumber space $P_{lm}(\mathbf{k},t,t_0)$. Equation (2.3) becomes then

$$R_{lm}(\mathbf{x},t,\mathbf{x}_0,t_0) = \int \mathrm{d}^3k \;\; P_{lm}(\mathbf{k},t,t_0)\,\mathrm{e}^{\mathrm{i}\mathbf{k}\cdot(\mathbf{x}-\mathbf{x}_0)}. \tag{2.5}$$

By setting $t_0 = 0$ and $\mathbf{x}_0 = 0$, we have

$$P_{lm}(\mathbf{k},t) = \langle \delta B_l(\mathbf{k},t)\delta B_m^*(\mathbf{k},0)\rangle \tag{2.6}$$

and, thus

$$R_{lm}(\mathbf{x},t) = \int \mathrm{d}^3k \;\; P_{lm}(\mathbf{k},t)\,\mathrm{e}^{\mathrm{i}\mathbf{k}\cdot\mathbf{x}}. \tag{2.7}$$

To proceed, we assume the same temporal behavior of all tensor components

$$P_{lm}(\mathbf{k},t) = P_{lm}(\mathbf{k},0)\;\Gamma(\mathbf{k},t), \tag{2.8}$$

where we used the dynamical correlation function $\Gamma(\mathbf{k},t)$ and the magnetostatic correlation tensor $P_{lm}(\mathbf{k},0) \equiv P_{lm}(\mathbf{k})$. In the following, we discuss several forms of the magnetostatic tensor and, in Sect. 2.5, different forms of $\Gamma(\mathbf{k},t)$.

2.1.1 The Isotropic Correlation Tensor

For our first considerations, we assume isotropic turbulence. It can be demonstrated (e.g., Batchelor 1953) that the general form of an isotropic (rank-2) tensor is

$$P_{lm}(\mathbf{k}) = A(k)\delta_{lm} + B(k)k_l k_m + C(k)\sum_n \varepsilon_{lmn}k_n. \tag{2.9}$$

From the Maxwell equations, we know the solenoidal constraint

$$\sum_i \partial_i \delta B_i(\mathbf{x}) = 0 \Rightarrow \sum_i k_i \delta B_i(\mathbf{k}) = 0 \tag{2.10}$$

and, therefore, we have

$$\sum_{l,m} \langle k_l \delta B_l k_m \delta B_m^*\rangle = \sum_{l,m} k_l k_m P_{lm} = 0. \tag{2.11}$$

By inserting (2.9) into this condition, we find

$$\begin{aligned}0 &= A(k)\sum_{l,m} k_l k_m \delta_{lm} + B(k)\sum_{l,m} k_l^2 k_m^2 + C(k)\sum_{l,m,n} \varepsilon_{lmn} k_l k_m k_n,\\ &= A(k)k^2 + B(k)k^4,\end{aligned} \tag{2.12}$$

and thus

$$B(k) = -\frac{A(k)}{k^2}. \tag{2.13}$$

Therefore, we can write (2.9) as

$$P_{lm}(\mathbf{k}) = A(k)\left(\delta_{lm} - \frac{k_l k_m}{k^2}\right) + C(k)\sum_n \varepsilon_{lmn} k_n. \tag{2.14}$$

In addition, we have

$$P_{lm}(\mathbf{k}) = \langle \delta B_l \delta B_m^* \rangle = \langle \delta B_l^* \delta B_m \rangle^* = \langle \delta B_m \delta B_l^* \rangle^* = P_{ml}^*(\mathbf{k}). \tag{2.15}$$

This condition can be used to obtain further information about the functions $A(k)$ and $C(k)$:

$$\begin{aligned}P_{ml}^* &= A^*(k)\left(\delta_{ml} - \frac{k_m k_l}{k^2}\right) + C^*(k)\sum_n \varepsilon_{mln} k_n,\\ &= A^*(k)\left(\delta_{lm} - \frac{k_l k_m}{k^2}\right) - C^*(k)\sum_n \varepsilon_{lmn} k_n,\\ &\overset{!}{=} A(k)\left(\delta_{lm} - \frac{k_l k_m}{k^2}\right) + C(k)\sum_n \varepsilon_{lmn} k_n.\end{aligned} \tag{2.16}$$

Therefore, $A(k) = A(k)^*$ and the function $A(k)$ must be real. Furthermore, we find $C(k) = -C(k)^*$ and the function $C(k)$ must be imaginary. Without loosing generality, we can set

$$C(k) = \mathrm{i}A(k)\frac{\sigma(k)}{k} \tag{2.17}$$

to find

$$P_{lm}(\mathbf{k}) = A(k)\left[\delta_{lm} - \frac{k_l k_m}{k^2} + \mathrm{i}\sigma(k)\sum_n \varepsilon_{lmn}\frac{k_n}{k}\right], \tag{2.18}$$

where $A(k)$ and $\sigma(k)$ are real numbers. $\sigma(k)$ is called the magnetic helicity. Both functions have to be replaced by appropriate models.

2.1.2 Axisymmetric Turbulence and Vanishing Magnetic Helicity

In the previous paragraph, a simple form of the correlation tensor has been derived. Although this form was derived under the assumption of isotropic turbulence, Matthaeus and Smith (1981) demonstrated that this form is also valid for axisymmetric (with respect to the z-axis) turbulence:

$$P_{lm}(\mathbf{k}) = A(k_\parallel, k_\perp)\left[\delta_{lm} - \frac{k_l k_m}{k^2} + \mathrm{i}\sigma(k_\parallel, k_\perp)\sum_n \varepsilon_{lmn}\frac{k_n}{k}\right]. \tag{2.19}$$

In our case, the symmetry axis has to be identified with the direction of the uniform mean magnetic field. As for most applications (e.g., plasma containment devices, interplanetary medium), the condition of axisymmetry should be well satisfied, in this book, (2.19) is used for turbulence theory considerations and for cosmic ray transport theory applications. Furthermore, we assume in all investigations that the magnetic helicity can be neglected and that the parallel component of the turbulent fields is zero or negligibly small ($\delta B_z = 0$). In this case, the correlation tensor can be further simplified:

$$P_{lm}(\mathbf{k}) = A(k_\parallel, k_\perp)\left(\delta_{lm} - \frac{k_l k_m}{k^2}\right), \quad l, m = x, y \tag{2.20}$$

and $P_{lz} = P_{zm} = 0$. This form of the correlation tensor is valid if

- The turbulence is axisymmetric with respect to the mean field $\mathbf{B}_0$
- The magnetic helicity can be neglected ($\sigma(k_\parallel, k_\perp) = 0$)
- We have vanishing or negligibly small parallel turbulent fields ($\delta B_z = 0$)

The function $A(k_\parallel, k_\perp)$ is controlled by two turbulence properties:

1. *The turbulence geometry:* the geometry describes how $A(k_\parallel, k_\perp)$ depends on the direction of the wave vector. In this book, three specific models for the turbulence geometry are considered:

   (a) *The slab (or one-dimensional) model:* Here, we assume that the turbulent fields depend only on the z-coordinate

   $$\delta B_i^{\text{slab}}(\mathbf{x}) = \delta B_i^{\text{slab}}(z) \tag{2.21}$$

   and thus the function $A(k_\parallel, k_\perp)$ used in (2.20) has the form

   $$A^{\text{slab}}(k_\parallel, k_\perp) = g^{\text{slab}}(k_\parallel)\frac{\delta(k_\perp)}{k_\perp}. \tag{2.22}$$

   In this model, the wave vectors are aligned parallel to the mean field: $\mathbf{k} \parallel \mathbf{B}_0$. In Fig. 2.1 the field lines of the slab model are shown.

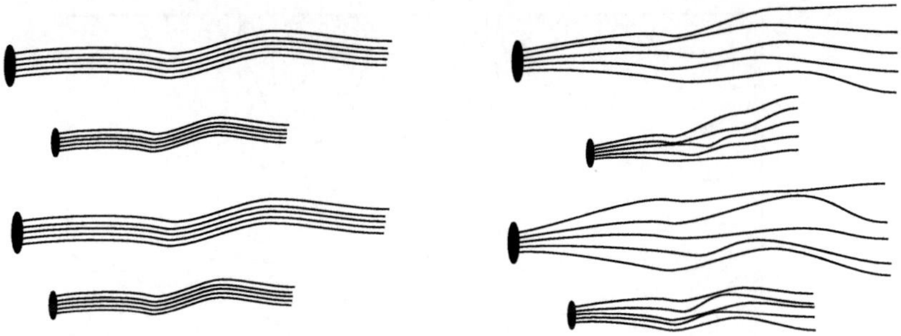

Fig. 2.1 The field lines of the slab model (*left panel*) and the slab/2D composite model (*right panel*)

(b) *The 2D (or perpendicular) model:* Here, the turbulent fields depend only on the perpendicular coordinates (x, y)

$$\delta B_i^{2D}(\mathbf{x}) = \delta B_i^{2D}(x, y) \tag{2.23}$$

and thus

$$A^{2D}(k_\parallel, k_\perp) = g^{2D}(k_\perp)\frac{\delta(k_\parallel)}{k_\perp}. \tag{2.24}$$

In this model, the wave vectors are aligned perpendicular to the mean field: $\mathbf{k} \perp \mathbf{B}_0$, and are, therefore, in a two-dimensional (2D) plane. It should be noted, however, that also 2D models with $\delta B_z \neq 0$ can be formulated. In this book, we consider only a 2D model with $\delta B_z = 0$, which could be called a *full 2D* model because $\mathbf{k} \perp \mathbf{B}_0$ and $\delta\mathbf{B} \perp \mathbf{B}_0$.

(c) *The slab/2D composite (or two-component) model:* In reality, the turbulent fields can depend on all three spatial coordinates. A quasi-three-dimensional model is the so-called slab/2D composite model, where we assume a superposition of the slab and the 2D model: $\delta B_i^{\mathrm{comp}}(x, y, z) = \delta B_i^{\mathrm{slab}}(z) + \delta B_i^{2D}(x, y)$. Because of $\langle \delta B_i^{\mathrm{slab}}(z) \delta B_j^{*,2D}(x, y) \rangle = 0$, the correlation tensor has the form

$$P_{lm}^{\mathrm{comp}}(\mathbf{k}) = P_{lm}^{\mathrm{slab}}(\mathbf{k}) + P_{lm}^{2D}(\mathbf{k}). \tag{2.25}$$

In the composite model, the total strength of the fluctuations is $\delta B^2 = \delta B_{\mathrm{slab}}^2 + \delta B_{2D}^2$. In Fig. 2.1, the field lines of the two-component model are compared with the field lines of the slab model. The composite model is often used to model solar wind turbulence. In the context of solar wind turbulence, the 2D model was originally introduced by Matthaeus et al. (1990). In Fig. 2.2, a contour plot of the correlation function of solar wind fluctuations is shown. This plot gives rise to the assumption that the slab/2D model is a good approximation for solar wind turbulence. Later (e.g., Bieber et al. 1996) it was demonstrated that 20% slab/80% 2D should be realistic in the solar wind at 1 AU heliocentric distance.

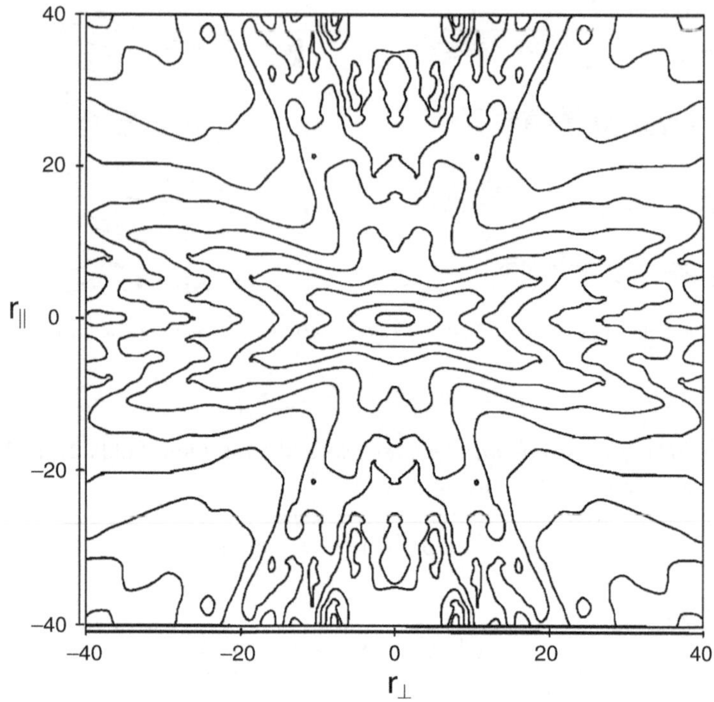

Fig. 2.2 Shown is a contour plot of the magnetic correlation function of solar wind fluctuations as a function of distance parallel and perpendicular to the mean magnetic field. The separations in $r_\parallel$ and $r_\perp$ are in units of 10^{10}cm. Reprinted with permission from The American Geophysical Union – Matthaeus et al. (1990)

For completeness, it should be noted, however, that other models have been considered previously by other authors. For example, Fisk (1974) and Bieber et al. (1988) applied an isotropic model for particle propagation studies. A generalization of the isotropic model is the anisotropic model used by Spangler (1991) to investigate the heating of the interstellar medium. Lerche and Schlickeiser (2001) and Teufel et al. (2003) used the same model to consider particle transport in magnetohydrodynamic turbulence. A theoretical approach to derive analytic forms for the correlation tensor has been proposed by Goldreich and Sridhar (1995).

2. *The wave spectrum:* The wave spectrum describes the wavenumber dependence of $A(k_\parallel, k_\perp)$. In the slab model, the spectrum is described by the function $g^{\text{slab}}(k_\parallel)$, and in the 2D model by $g^{2\text{D}}(k_\perp)$. In the following, we choose an appropriate model for the spectrum, which is in agreement with solar wind observations (see Fig. 2.3) obtained by the Helios 2 mission. Obviously, the observed spectrum consists of three ranges:

   - For small wavenumbers or large scales, we find that the spectrum is $\sim k^{-1}$. At such large spatial turbulence scales, the turbulence gains energy; therefore,

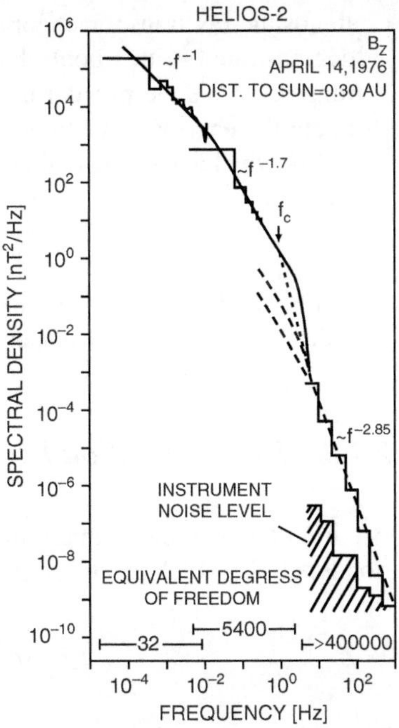

Fig. 2.3 Observed wave spectrum in the heliosphere obtained by the Helios 2 mission. Clearly, we can see the three ranges of the spectrum: for small wavenumbers, we find $\sim k^{-1}$ (energy range); for medium wavenumbers, we find a Kolmogorov-spectrum $\sim k^{-5/3}$ (inertial range); and for large wavenumbers, we find a steep spectrum $\sim k^{-3}$ (dissipation range). Reprinted with permission from NASA Conference Publication – Denskat and Neubauer (1982)

this part of the spectrum is called the energy range. To define mathematically the energy range, we introduce the bendover or turnover scale l_i:

$$g^i\,(k_i \leq l_i^{-1}) = \text{energy range of the spectrum.} \tag{2.26}$$

Here, i stands for the different turbulence geometries (e.g., $i = slab$ for the slab model).

- For medium scales or wavenumbers, the turbulence energy is transfered from large to small scales. Therefore, this part of the spectrum is often called the inertial range. Here, we find for the spectrum $\sim k^{-5/3}$, in agreement with the turbulence theory of Kolmogorov (1941). Mathematically, we define the inertial range by introducing the dissipation wavenumber $k_{d,i}$:

$$g^i\,(l_i^{-1} \leq k_i \leq k_{d,i}) = \text{inertial range of the spectrum.} \tag{2.27}$$

- For high wavenumbers or small scales, the turbulence loses energy through dissipation. Therefore, the high wavenumber part of the spectrum is called the dissipation range, where we have approximately $\sim k^{-3}$. The mathematical definition is

$$g^i\,(k_{d,i} \leq k_i) = \text{dissipation range of the spectrum.} \tag{2.28}$$

In cosmic ray transport theory, we can often neglect the dissipation range. It is demonstrated in this book that only for small energy particles, the dissipation range controls the parallel mean free path. Furthermore, it is useful to approximate the energy range by a constant. Therefore, as suggested by Bieber et al. (1994), the kappa distribution

$$g^i(k_i) \sim (1 + k_i^2 l_i^2)^{-\nu} \tag{2.29}$$

can be used to represent heliospheric turbulence with the inertial range spectral index $s = 2\nu$. A modified spectrum, with dissipation range, is used in Chap. 8.

2.1.3 The Correlation Length

An important parameter in turbulence and transport theory is the so-called correlation length l_c, which can be defined as (Matthaeus and Goldstein 1982)

$$l_{c,i,j}\delta B_j^2 = \int_0^\infty \mathrm{d}x_i \; R_{jj}(\mathbf{x}, 0). \tag{2.30}$$

It is a characteristic length scale for the spatial decorrelation of the turbulence. The correlation length is defined so that the area $l_{c,i,j}\delta B_j^2$ is equal to the area under the correlation function (see Fig. 2.4). In the following, the correlation functions are calculated, as well as the correlation lengths for different turbulence models.

2.2 The Magnetostatic Slab Model

In magnetostatic turbulence, the dynamical correlation function (see (2.8)) is replaced by $\Gamma(\mathbf{k}, t) = 1$. For slab geometry, we assume that the turbulent magnetic fields vary only in the parallel direction ($\delta B_i(\mathbf{x}) = \delta B_i(z)$) and the correlation functions have the form

$$R_{lm}^{\mathrm{slab}}(z) = \langle \delta B_l(z) \delta B_m^*(0) \rangle, \tag{2.31}$$

where we used $z(0) = 0$ due to the assumption of homogeneous turbulence. Or, in wavenumber space (see (2.20) with (2.22))

$$P_{lm}^{\mathrm{slab}}(\mathbf{k}) = g^{\mathrm{slab}}(k_\parallel) \frac{\delta(k_\perp)}{k_\perp} \begin{cases} \delta_{lm} \text{ if } l, m = x, y \\ 0 \;\text{ if } l \text{ or } m = z, \end{cases} \tag{2.32}$$

with the slab wave spectrum $g^{\mathrm{slab}}(k_\parallel)$. In Sect. 2.1, it is argued that the following form for the spectrum is appropriate (at least for solar wind turbulence)

$$g^{\mathrm{slab}}(k_{\parallel}) = \frac{C(\nu)}{2\pi} l_{\mathrm{slab}}\, \delta B_{\mathrm{slab}}^{2} (1 + k_{\parallel}^{2} l_{\mathrm{slab}}^{2})^{-\nu}. \tag{2.33}$$

Here we used the slab bendover scale l_{slab}, the strength of the slab magnetic field $\delta B_{\mathrm{slab}}^{2}$, and the inertial range spectral index 2ν. From the normalization condition of the spectrum

$$\delta B_{\mathrm{slab}}^{2} = \delta B_{x}^{2} + \delta B_{y}^{2} = R_{xx}(0) + R_{yy}(0) = \int \mathrm{d}^{3}k\, \left[P_{xx}(\mathbf{k}) + P_{yy}(\mathbf{k})\right], \tag{2.34}$$

we find

$$\delta B_{\mathrm{slab}}^{2} = 2 \int \mathrm{d}^{3}k\, g^{\mathrm{slab}}(k_{\parallel}) \frac{\delta(k_{\perp})}{k_{\perp}} = 8\pi \int_{0}^{\infty} \mathrm{d}k_{\parallel}\, g^{\mathrm{slab}}(k_{\parallel}). \tag{2.35}$$

With (2.33), it follows

$$\frac{1}{C(\nu)} = 4 l_{\mathrm{slab}} \int_{0}^{\infty} \mathrm{d}k_{\parallel}\, (1 + k_{\parallel}^{2} l_{\mathrm{slab}}^{2})^{-\nu}. \tag{2.36}$$

The integral can be solved (see Gradshteyn and Ryzhik 2000) and we obtain

$$C(\nu) = \frac{1}{2\sqrt{\pi}} \frac{\Gamma(\nu)}{\Gamma(\nu - 1/2)}, \tag{2.37}$$

where we used the Gamma function $\Gamma(x)$. In the following, we deduce analytic forms for the correlation function and for the correlation length by using the spectrum defined in (2.33).

2.2.1 The Slab Correlation Function

By combining (2.33) with (2.32) and (2.7), the slab correlation function can be calculated:

$$\begin{aligned} R_{xx}^{\mathrm{slab}}(z) = \langle \delta B_{x}(z) \delta B_{x}^{*}(0) \rangle &= \int \mathrm{d}^{3}k\, P_{xx}^{\mathrm{slab}}(\mathbf{k}) \cos(k_{\parallel} z), \\ &= 4\pi \int_{0}^{\infty} \mathrm{d}k_{\parallel}\, g^{\mathrm{slab}}(k_{\parallel}) \cos(k_{\parallel} z), \\ &= 2C(\nu) \delta B_{\mathrm{slab}}^{2} \int_{0}^{\infty} \mathrm{d}x\, (1 + x^{2})^{-\nu} \cos(ax). \end{aligned} \tag{2.38}$$

In the last step, we applied the integral transformation $x = k_{\parallel} l_{\mathrm{slab}}$ and we used the parameter $a = z/l_{\mathrm{slab}}$. The last integral in (2.38) can be solved (see Gradshteyn and

Ryzhik 2000, Eq. (3.771.2)):

$$\int_0^\infty \mathrm{d}x \; (1+x^2)^{-\nu} \cos(ax) = \frac{\sqrt{\pi}}{\Gamma(\nu)} \left(\frac{2}{a}\right)^{1/2-\nu} K_{\nu-1/2}(a), \tag{2.39}$$

where we used the modified Bessel function of imaginary argument, $K_\lambda(z)$. We define the perpendicular correlation function as $R_\perp = R_{xx} + R_{yy}$ to get

$$R_\perp^{\text{slab}}(z) = \frac{2\delta B_{\text{slab}}^2}{\Gamma(\nu - 1/2)} \left(\frac{2l_{\text{slab}}}{z}\right)^{1/2-\nu} K_{\nu-1/2}\left(\frac{z}{l_{\text{slab}}}\right). \tag{2.40}$$

To simplify this result, we consider two extreme cases:

- For $z \ll l_{\text{slab}}$, we can use (Abramowitz and Stegun 1974, Eq. (9.6.9))

$$K_{\nu-1/2}\,(z \ll l_{\text{slab}}) \approx \frac{1}{2}\Gamma(\nu - 1/2) \left(\frac{2l_{\text{slab}}}{z}\right)^{\nu-1/2} \tag{2.41}$$

  (if $\nu > 1/2$) to find the expected result

$$R_\perp^{\text{slab}}(z \ll l_{\text{slab}}) = \delta B_{\text{slab}}^2. \tag{2.42}$$

- More interesting is the case $z \gg l_{\text{slab}}$. Here we can use (Abramowitz and Stegun 1974, Eq. (9.7.2))

$$K_{\nu-1/2}\,(z \gg l_{\text{slab}}) \approx \sqrt{\frac{\pi l_{\text{slab}}}{2z}}\, \mathrm{e}^{-z/l_{\text{slab}}} \tag{2.43}$$

  and thus

$$R_\perp^{\text{slab}}(z \gg l_{\text{slab}}) = \frac{\sqrt{\pi}}{\Gamma(\nu - 1/2)} \delta B_{\text{slab}}^2 \left(\frac{2l_{\text{slab}}}{z}\right)^{1-\nu} \mathrm{e}^{-z/l_{\text{slab}}}. \tag{2.44}$$

Obviously, the correlation function decays exponentially. The characteristic length scale for the spatial decorrelation of turbulence is the slab bendover scale l_{slab}. Equation (2.38) is also examined numerically. In Fig. 2.4, the results are compared with the analytical approximations.

2.2.2 *The Slab Correlation Length*

Here, we calculate the slab correlation length $l_{\text{c,slab}}$ (see (2.30))

$$l_{\text{c,slab}} \delta B_{\text{slab}}^2 = \int_0^\infty \mathrm{d}z \; R_\perp^{\text{slab}}(z). \tag{2.45}$$

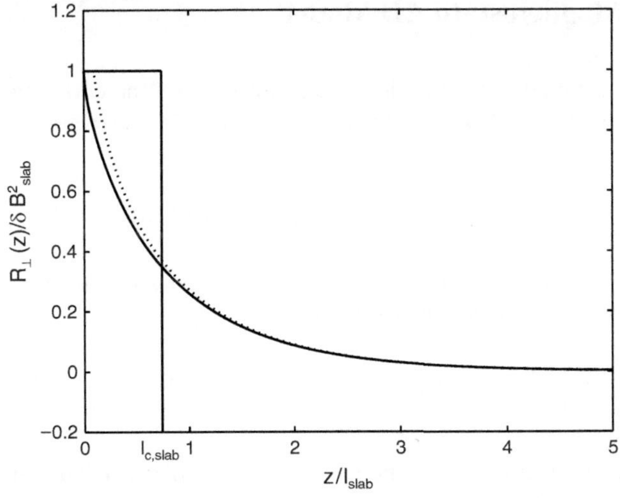

Fig. 2.4 The slab-correlation function $R_\perp/\delta B_{\rm slab}^2$ for $\nu = 5/6$ as a function of the parallel position $z/l_{\rm slab}$. Shown are numerical results (*solid line*) and the analytical approximation for $z \gg l_{\rm slab}$ (*dotted line*). For $z = 0$ we have $R_\perp = \delta B_{\rm slab}^2$. Also, the slab-correlation-length $l_{\rm c,slab}$ is shown; note that the area under the box is the same as the area under the correlation function

By using (2.32) and (2.7), we find

$$\begin{aligned} l_{\rm c,slab}\delta B_{\rm slab}^2 &= 2\pi \int_{-\infty}^{+\infty} {\rm d}k_\parallel\ g^{\rm slab}(k_\parallel) \int_{-\infty}^{+\infty} {\rm d}z\, {\rm e}^{{\rm i}k_\parallel z}, \\ &= (2\pi)^2 \int_{-\infty}^{+\infty} {\rm d}k_\parallel\ g^{slab}(k_\parallel)\delta(k_\parallel), \\ &= (2\pi)^2 g^{slab}(0). \end{aligned} \tag{2.46}$$

With the spectrum from (2.33), we finally find

$$l_{\rm c,slab}\delta B_{\rm slab}^2 = \int_0^\infty {\rm d}z\ R_\perp^{\rm slab}(z) = 2\pi C(\nu) l_{\rm slab} B_{\rm slab}^2. \tag{2.47}$$

Obviously, the correlation length for the used spectrum is related to the slab bendover scale through

$$l_{\rm c,slab} = 2\pi C(\nu) l_{\rm slab}. \tag{2.48}$$

For $\nu = 5/6$, we have $C(\nu = 5/6) = 0.1188$ and thus $l_{\rm c,slab} \approx 0.75 l_{\rm slab}$. Both quantities are similar for the applied spectrum. In general, these two characteristic length scales of the turbulence are not to be similar nor must they be comparable (see Sect. 2.3).

2.3 The Magnetostatic 2D Model

For pure 2D geometry, the turbulence depends only on the coordinates in the perpendicular direction $\delta B_i(\mathbf{x}) = \delta B_i(x, y)$. Thus, the 2D correlation tensor has the form

$$R_{lm}^{\mathrm{2D,MS}}(x, y) = \langle \delta B_l(x, y) \delta B_m^*(0) \rangle, \tag{2.49}$$

where we used $x_0 = y_0 = 0$. By using the 2D model introduced in (2.24), the correlation tensor in wavenumber space is

$$P_{lm}^{\mathrm{2D}}(\mathbf{k}) = g^{\mathrm{2D}}(k_\perp) \frac{\delta(k_\parallel)}{k_\perp} \begin{cases} \delta_{lm} - \frac{k_l k_m}{k^2} & \text{if } l, m = x, y \\ 0 & \text{if } l \text{ or } m = z. \end{cases} \tag{2.50}$$

For the wave spectrum, we assume the same (normalized) form that was used for the slab model

$$g^{\mathrm{2D}}(k_\perp) = \frac{2C(\nu)}{\pi} l_{\mathrm{2D}} \delta B_{\mathrm{2D}}^2 (1 + k_\perp^2 l_{\mathrm{2D}}^2)^{-\nu}. \tag{2.51}$$

Here we used the strength of the 2D fluctuations δB_{2D}^2 and the 2D bendover scale l_{2D}. In the following, we calculate the 2D correlation function and the 2D correlation length.

2.3.1 The 2D Correlation Function

Here, we calculate the correlation function $R_{xx}(x, y)$ by repeating the steps of the previous paragraph. With

$$R_{xx}(x, y) = \int d^3k \; P_{xx}(\mathbf{k}) \, e^{i\mathbf{k}\cdot\mathbf{x}} = \int d^3k \; P_{xx}(\mathbf{k}) \, e^{ik_x x + ik_y y} \tag{2.52}$$

and by using cylindrical coordinates for the wave vector and for the position

$$\begin{aligned} k_x &= k_\perp \cos\Psi, \quad & x &= \rho \cos\Phi, \\ k_y &= k_\perp \sin\Psi, \quad & y &= \rho \sin\Phi, \end{aligned} \tag{2.53}$$

we find

$$\begin{aligned} R_{xx}^{\mathrm{2D}}(x, y) &= \int_0^\infty dk_\perp \; g^{\mathrm{2D}}(k_\perp) \int_0^{2\pi} d\Psi \; \sin^2\Psi \; e^{ik_\perp \rho(\cos\Phi \cos\Psi + \sin\Phi \sin\Psi)}, \\ &= \int_0^\infty dk_\perp \; g^{\mathrm{2D}}(k_\perp) \int_0^{2\pi} d\Psi \; \sin^2\Psi \; e^{ik_\perp \rho \cos(\Phi - \Psi)}. \end{aligned} \tag{2.54}$$

Such expressions can be simplified by using the formulas (see, e.g., Abramowitz and Stegun 1974)

$$e^{ix \sin(\alpha)} = \sum_{n=-\infty}^{\infty} J_n(x)\, e^{in\alpha}, \quad e^{ix \cos(\alpha)} = \sum_{n=-\infty}^{\infty} J_n(x)\, e^{in(\alpha+\pi/2)}. \tag{2.55}$$

With the latter relation, we find

$$R_{xx}^{2D}(x, y) = \int_0^{\infty} dk_\perp\, g^{2D}(k_\perp) \sum_{n=-\infty}^{\infty} J_n(k_\perp \rho) \int_0^{2\pi} d\Psi\, \sin^2\Psi\, e^{-in\Psi} e^{in(\Phi+\pi/2)} \tag{2.56}$$

and, analogously,

$$R_{yy}^{2D}(x, y) = \int_0^{\infty} dk_\perp\, g^{2D}(k_\perp) \sum_{n=-\infty}^{\infty} J_n(k_\perp \rho) \times \int_0^{2\pi} d\Psi\, \cos^2\Psi\, e^{-in\Psi} e^{in(\Phi+\pi/2)}. \tag{2.57}$$

Thus, we find for the combined perpendicular correlation function ($R_\perp = R_{xx} + R_{yy}$)

$$R_\perp^{2D}(x, y) = \int_0^{\infty} dk_\perp\, g^{2D}(k_\perp) \sum_{n=-\infty}^{\infty} J_n(k_\perp \rho) \int_0^{2\pi} d\Psi\, e^{-in\Psi} e^{in(\Phi+\pi/2)}. \tag{2.58}$$

With

$$\int_0^{2\pi} d\Psi\, e^{\pm in\Psi} = 2\pi \delta_{n0}, \tag{2.59}$$

where δ_{n0} denotes the Kronecker symbol, we obtain

$$R_\perp^{2D}(\rho) = 2\pi \int_0^{\infty} dk_\perp\, g^{2D}(k_\perp) J_0(k_\perp \rho). \tag{2.60}$$

With the integral transformation $x = k_\perp l_{2D}$, with $a = \rho / l_{2D}$ and by using the spectrum of (2.51), we have

$$R_\perp^{2D}(\rho) = 4C(\nu)\delta B_{2D}^2 \int_0^{\infty} dx\, (1 + x^2)^{-\nu} J_0(ax). \tag{2.61}$$

In the following we consider again two different limits:

- For $a = 0$ and with $J_0(0) = 1$, we find with

$$\int_0^{\infty} dx\, (1 + x^2)^{-\nu} = \frac{1}{4C(\nu)} \tag{2.62}$$

the expected result

$$R_{\perp}^{2D}(\rho = 0) = \delta B_{2D}^2. \tag{2.63}$$

- In the limit $a \to \infty$, the integral can be well approximated by

$$\begin{aligned}\int_0^\infty dx\,(1+x^2)^{-\nu} J_0(ax), &= \frac{1}{a}\int_0^\infty dy\,\left(1+\frac{y^2}{a^2}\right)^{-\nu} J_0(y),\\ &\approx \frac{1}{a}\int_0^\infty dy\, J_0(y)\\ &= \frac{1}{a}.\end{aligned} \tag{2.64}$$

Here we used the integral-transformation $y = ax$ and we have employed

$$\int_0^\infty dy\, J_0(y) = 1. \tag{2.65}$$

Thus we find

$$R_{\perp}(\rho \gg l_{2D}) = 4C(\nu)\delta B_{2D}^2 \frac{l_{2D}}{\rho}. \tag{2.66}$$

Similar to the slab model, the characteristic length scale for the spatial decorrelation of the turbulence is again the bendover scale (here l_{2D}). In Fig. 2.5, the 2D correlation function is shown. Although the wave spectra of slab and 2D are the same, the correlation functions are different: the 2D correlation function decays much slower with increasing distance.

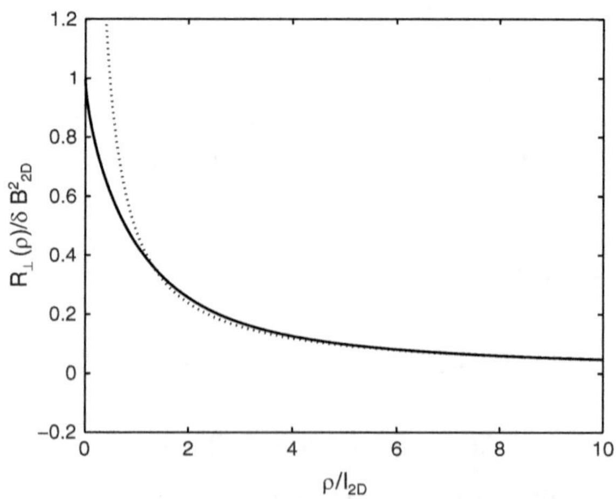

Fig. 2.5 The 2D correlation function $R_{\perp}/\delta B_{2D}^2$ for $\nu = 5/6$ as a function of the perpendicular position ρ/l_{2D}. Shown are numerical results (*solid line*) and the analytical approximation for $z \gg l_{2D}$ (*dotted line*). For $\rho = 0$ we have $R_{\perp} = \delta B_{2D}^2$

2.3.2 The Correlation Length for Pure 2D Turbulence

To perform the above calculations, we use a spectrum with a large but finite box-size L_{2D}:

$$\begin{aligned} l_{c,2D} &= \frac{1}{\delta B_{2D}^2} \int_0^\infty d\rho \; R_\perp(\rho), \\ &= 4C(\nu) \int_{x_{min}}^\infty dx \; (1+x^2)^{-\nu} \int_0^\infty d\rho \; J_0\left(\frac{x\rho}{l_{2D}}\right) \\ &= 4C(\nu) l_{2D} \int_{x_{min}}^\infty \frac{dx}{x} \; (1+x^2)^{-\nu} \end{aligned} \tag{2.67}$$

where we used (2.61), $x_{min} = l_{2D}/L_{2D}$, and (2.65). The remaining integral can be solved approximately

$$\begin{aligned} l_{c,2D} &\approx 4C(\nu) l_{2D} \left(\int_{x_{min}}^1 \frac{dx}{x} + \int_1^\infty dx \; x^{-2\nu-1} \right), \\ &\approx 4C(\nu) l_{2D} \left(\frac{1}{2\nu} + \ln \frac{L_{2D}}{l_{2D}} \right). \end{aligned} \tag{2.68}$$

The wave spectrum used here (2.51) is correctly normalized only if $L_{2D} \gg l_{2D}$. In the limit of an infinitely large box-size ($L_{2D} \to \infty$), we find an infinitely large correlation length ($l_{c,2D} \to \infty$).

2.3.3 The Vector Potential of Pure 2D Turbulence

For certain transport theory calculations in pure 2D turbulence (see next chapter), it is convenient to apply a vector potential approach. For pure 2D fluctuations, the correlation tensor components have the form

$$P_{lm}(\mathbf{k}) = A_{zz}(\mathbf{k}) \left(\delta_{lm} k_\perp^2 - k_l k_m \right) \quad l,m = x,y, \tag{2.69}$$

with (see (2.50))

$$A_{zz}(\mathbf{k}) := g^{2D}(k_\perp) \frac{\delta(k_\parallel)}{k_\perp^3}. \tag{2.70}$$

The vector potential $\delta\mathbf{A}(\mathbf{x})$ is related to the magnetic fields via

$$\delta\mathbf{B} = \nabla \times \delta\mathbf{A}, \tag{2.71}$$

and thus

$$\delta B_x(\mathbf{x}) = \partial_y \delta A_z(\mathbf{x}) - \partial_z \delta A_y(\mathbf{x}) = \partial_y \delta A_z(x, y),$$
$$\delta B_y(\mathbf{x}) = \partial_z \delta A_x(\mathbf{x}) - \partial_x \delta A_z(\mathbf{x}) = -\partial_x \delta A_z(x, y), \tag{2.72}$$

where we used $\delta\mathbf{A}(\mathbf{x}) = \delta\mathbf{A}(x, y)$ for pure 2D turbulence. In Fourier space, this becomes

$$\delta B_x(k_x, k_y) = \mathrm{i}k_y \delta A_z(k_x, k_y),$$
$$\delta B_y(k_x, k_y) = -\mathrm{i}k_x \delta A_z(k_x, k_y), \tag{2.73}$$

and thus

$$P_{lm}(\mathbf{k}) = \langle \delta B_l(k_x, k_y) \delta B_m^*(k_x, k_y) \rangle$$
$$= \langle \delta A_z(k_x, k_y) \delta A_z^*(k_x, k_y) \rangle \left[\delta_{lm} k_\perp^2 - k_l k_m \right]. \tag{2.74}$$

By comparing (2.74) with (2.69), the function $A_{zz}(\mathbf{k})$ can be identified with the correlation function of the vector potential:

$$A_{zz}(k_x, k_y) = \langle \delta A_z(k_x, k_y) \delta A_z^*(k_x, k_y) \rangle. \tag{2.75}$$

In Chap. 3, these results will be used to describe cosmic ray pitch-angle diffusion in pure 2D turbulence.

2.4 Linear and Nonlinear Theories for Stochastic Field Line Wandering

In turbulence theory, an important quantity is the field line diffusion coefficient, which has been discussed in several articles (e.g., Jokipii and Parker 1969; Matthaeus et al. 1995; Ruffolo et al. 2004, 2006; Shalchi and Kourakis 2007). In this section, we discuss different approaches for describing field line random walk analytically in magnetostatic turbulence. The equation for the field lines $\mathbf{x} = (x(z), y(z), z)$ can be written as

$$\mathrm{d}x \; \delta B_z(\mathbf{x}) = \mathrm{d}z \; \delta B_x(\mathbf{x}) \tag{2.76}$$

if we adopt purely magnetic fields with vanishing z component ($\delta B_z = 0$). Furthermore, we assume that the mean field (or background field) is aligned parallel to the z-axis ($\mathbf{B}_0 = B_0 \mathbf{e}_z$). The solution of this equation are the field lines: $x = x(z)$. A similar equation can be found for the y component. The displacement in the x direction can be written as the following integral equation:

$$\Delta x(z) = \frac{1}{B_0} \int_0^z \mathrm{d}z' \; \delta B_x(\mathbf{x}(z')) \tag{2.77}$$

and, thus, we find for the mean square displacement

$$\left\langle (\Delta x(z))^2 \right\rangle = \frac{1}{B_0^2} \int_0^z \mathrm{d}z' \int_0^z \mathrm{d}z'' \left\langle \delta B_x(\mathbf{x}(z')) \delta B_x^*(\mathbf{x}(z'')) \right\rangle . \qquad (2.78)$$

By applying a Fourier representation for the magnetic fields, we can easily derive

$$\left\langle (\Delta x(z))^2 \right\rangle = \frac{1}{B_0^2} \int \mathrm{d}^3k \int \mathrm{d}^3k' \int_0^z \mathrm{d}z' \int_0^z \mathrm{d}z'' \left\langle \delta B_x(\mathbf{k}) \delta B_x^*(\mathbf{k}')\, \mathrm{e}^{\mathrm{i}\mathbf{k}\cdot\mathbf{x}(z') - \mathrm{i}\mathbf{k}'\cdot\mathbf{x}(z'')} \right\rangle . \qquad (2.79)$$

To proceed, we must distinguish two different cases:

- *The slab model:* Here the exponential function in the brackets $\langle \dots \rangle$ depends only on the variable z. In this case, we have $\mathbf{k} \cdot \mathbf{x}(z) = k_{\parallel} z$, and the expression of (2.79) can be evaluated without using further assumptions or models.
- *Nonslab models:* Here the exponential function in the brackets $\langle \dots \rangle$ can depend on all components of the field line vector $\mathbf{x}(z) = (x(z), y(z), z)$. In this case, the field line equation (2.79) is a nonlinear equation. Thus, a quasilinear or a nonlinear theory has to be applied to describe field line wandering in nonslab models such as the two-component model.

Before we investigate these two cases, we explore the limit $z \to 0$.

2.4.1 The Initial Free-Streaming Regime

In the limit $z \to 0$, we also expect $x, y \to 0$ and thus we can derive from (2.79)

$$\left\langle (\Delta x(z))^2 \right\rangle = \frac{1}{B_0^2} \int \mathrm{d}^3k \int \mathrm{d}^3k' \int_0^z \mathrm{d}z' \int_0^z \mathrm{d}z'' \left\langle \delta B_x(\mathbf{k}) \delta B_x^*(\mathbf{k}') \right\rangle . \qquad (2.80)$$

By using (2.4)

$$\left\langle \delta B_x(\mathbf{k}) \delta B_x^*(\mathbf{k}') \right\rangle = P_{xx}(\mathbf{k}) \delta\left(\mathbf{k} - \mathbf{k}'\right) \qquad (2.81)$$

this can be simplified to

$$\left\langle (\Delta x(z))^2 \right\rangle = \frac{z^2}{B_0^2} \int \mathrm{d}^3k \; P_{xx}(\mathbf{k}) = z^2 \frac{\delta B_x^2}{B_0^2}, \qquad (2.82)$$

and we find a strong superdiffusive behavior for small z, independent of the turbulence model. In the following, we refer to this limit as the initial free-streaming regime of the field lines.

2.4.2 Field Line Random Walk for Slab Turbulence

In (2.79), we deduced a general form of the field line mean square displacement. For pure slab geometry, this becomes

$$\left\langle (\Delta x(z))^2 \right\rangle = \frac{1}{B_0^2} \int d^3k \int d^3k' \left\langle \delta B_x(\mathbf{k}) \delta B_x^*(\mathbf{k}') \right\rangle \int_0^z dz' \int_0^z dz'' \, e^{ik_\parallel z' - k_\parallel' z''}. \tag{2.83}$$

By applying again (2.81) and by using (2.32) for the slab model, we find for the mean square deviation of the field lines

$$\left\langle (\Delta x(z))^2 \right\rangle = \frac{2\pi}{B_0^2} \int_{-\infty}^{\infty} dk_\parallel \, g^{\mathrm{slab}}(k_\parallel) \int_0^z dz' \int_0^z dz'' \, e^{ik_\parallel (z' - z'')}. \tag{2.84}$$

The two integrals over z' and z'' can easily be solved and we find

$$\left\langle (\Delta x(z))^2 \right\rangle = \frac{4\pi}{B_0^2} \int_{-\infty}^{\infty} dk_\parallel \, g^{\mathrm{slab}}(k_\parallel) \frac{1 - \cos\left(k_\parallel z\right)}{k_\parallel^2}. \tag{2.85}$$

We are interested in the stable regime of the variable z, and thus we can use (e.g., Jaekel and Schlickeiser 1992)

$$\left[\frac{1 - \cos(a\xi)}{\xi^2} \right]_{a \to \infty} = |a| \pi \delta(\xi) \tag{2.86}$$

to obtain

$$\begin{aligned} \left\langle (\Delta x(z))^2 \right\rangle &= \frac{4\pi^2}{B_0^2} |z| \int_{-\infty}^{\infty} dk_\parallel \, g^{\mathrm{slab}}(k_\parallel) \delta(k_\parallel), \\ &= \frac{4\pi^2}{B_0^2} |z| g^{\mathrm{slab}}(0). \end{aligned} \tag{2.87}$$

With the spectrum from (2.33), this becomes

$$\left\langle (\Delta x(z))^2 \right\rangle = 2\pi C(\nu) l_{\mathrm{slab}} |z| \frac{\delta B_{\mathrm{slab}}^2}{B_0^2} = 2\kappa_{\mathrm{FL}} |z|. \tag{2.88}$$

Obviously, we find $\left\langle (\Delta x(z))^2 \right\rangle \sim |z|$, which has to be interpreted as normal (Markovian) diffusion. On the right-hand side of (2.88), we introduced the field line diffusion coefficient

$$\kappa_{\mathrm{FL}} = \pi C(\nu) l_{\mathrm{slab}} \frac{\delta B_{\mathrm{slab}}^2}{B_0^2} = \frac{1}{2} l_{\mathrm{c,slab}} \frac{\delta B_{\mathrm{slab}}^2}{B_0^2}. \tag{2.89}$$

Here, we applied the relation $l_{c,\text{slab}} = 2\pi C(\nu) l_{\text{slab}}$ to substitute the slab bendover scale l_{slab} by the slab correlation length of the turbulence $l_{c,\text{slab}}$. In the following, we consider three different approaches to evaluate (2.79) for nonslab models.

2.4.3 *Quasilinear Theory of Field Line Random Walk*

Equation (2.79) is the exact form of the mean square deviation of the field lines. For nonslab models, this formula is a nonlinear equation for field line random walk. Within the quasilinear theory (QLT, Jokipii 1966) of field line wandering, we replace the field lines on the right-hand side of (2.79) by the unperturbed field lines (in this case we have $x(z) = y(z) = 0$) and thus

$$\left\langle (\Delta x(z))^2 \right\rangle = \frac{1}{B_0^2} \int d^3k \; P_{xx}(\mathbf{k}) \int_0^z dz' \int_0^z dz'' \; e^{ik_\parallel (z'-z'')} \tag{2.90}$$

where we again applied (2.81). The integral over z' and z'' can easily be solved and we find

$$\left\langle (\Delta x(z))^2 \right\rangle = \frac{2}{B_0^2} \int d^3k \; P_{xx}(\mathbf{k}) \frac{1 - \cos\left(k_\parallel z\right)}{k_\parallel^2}. \tag{2.91}$$

For pure slab geometry, this result reduces to (2.85). Thus, for pure slab turbulence, QLT for field line random walk is exact. For slab/2D composite turbulence, however, one gets

$$\left\langle (\Delta x(z))^2 \right\rangle = 2|z|\kappa_{\text{FL}} + \frac{z^2}{B_0^2} \int d^3k \; P_{xx}^{2\text{D}}(\mathbf{k}) \approx \frac{z^2}{2} \frac{\delta B_{2\text{D}}^2}{B_0^2} \tag{2.92}$$

in the limit $|\, z \,| \to \infty$. Here we used the slab field line diffusion coefficient κ_{FL} of (2.89). This result corresponds to a free-streaming behavior of the field lines. Obviously, QLT is not appropriate to describe field line wandering in the two-component model.

2.4.4 *The Nonlinear Approach for Field Line Random Walk*

A nonlinear theory for field line random walk was proposed by Shalchi and Kourakis (2007), which is a generalization of the Matthaeus et al. (1995) theory (see next paragraph). Within a nonlinear formulation, we have to apply Corrsin's independence hypothesis (Corrsin 1959; Salu and Montgomery 1977; McComb 1990; Matthaeus et al. 2003) onto (2.79) to obtain

$$\left\langle \delta B_x(\mathbf{k}) \delta B_x^*(\mathbf{k}') \, e^{i\mathbf{k}\cdot\mathbf{x}(z') - i\mathbf{k}'\cdot\mathbf{x}(z'')} \right\rangle = \left\langle \delta B_x(\mathbf{k}) \delta B_x^*(\mathbf{k}') \right\rangle \left\langle e^{i\mathbf{k}\cdot\mathbf{x}(z') - i\mathbf{k}'\cdot\mathbf{x}(z'')} \right\rangle. \tag{2.93}$$

To proceed, we apply (2.81) to find

$$\left\langle (\Delta x(z))^2 \right\rangle = \frac{1}{B_0^2} \Re\mathfrak{e} \int d^3k \; P_{xx}(\mathbf{k}) \int_0^z dz' \int_0^z dz'' \left\langle e^{i\mathbf{k}\cdot[\mathbf{x}(z')-\mathbf{x}(z'')]} \right\rangle. \quad (2.94)$$

For homogeneous turbulence, the term in the brackets $\langle \ldots \rangle$ depends only on $|z'-z''|$ and we can use

$$\int_0^z dz' \int_0^z dz'' \left\langle e^{i\mathbf{k}\cdot[\mathbf{x}(z')-\mathbf{x}(z'')]} \right\rangle = 2\int_0^z dz' \; (z-z') \left\langle e^{i\mathbf{k}\cdot\Delta\mathbf{x}(z')} \right\rangle. \quad (2.95)$$

To deduce (2.95), we applied the same steps as used in Chap. 1 (1.22)–(1.27). Hence, (2.94) becomes

$$\left\langle (\Delta x(z))^2 \right\rangle = \frac{2}{B_0^2} \Re\mathfrak{e} \int d^3k \; P_{xx}(\mathbf{k}) \int_0^z dz' \; (z-z') \left\langle e^{i\mathbf{k}\cdot\Delta\mathbf{x}(z')} \right\rangle, \quad (2.96)$$

with $\Delta\mathbf{x}(z) = \mathbf{x}(z) - \mathbf{x}(0)$. To proceed, we assume a Gaussian statistics of the field lines[1]

$$\left\langle e^{i\mathbf{k}\cdot\Delta\mathbf{x}(z)} \right\rangle = e^{-\frac{1}{2}\langle(\Delta x)^2\rangle k_x^2 - \frac{1}{2}\langle(\Delta y)^2\rangle k_y^2 + ik_\parallel z}. \quad (2.97)$$

For axisymmetric turbulence, we have $\left\langle (\Delta x)^2 \right\rangle = \left\langle (\Delta y)^2 \right\rangle$, and thus (2.96) becomes

$$\left\langle (\Delta x(z))^2 \right\rangle = \frac{2}{B_0^2} \int d^3k \; P_{xx}(\mathbf{k}) \int_0^z dz' \; (z-z') \cos\left(k_\parallel z'\right) \, e^{-\frac{1}{2}\langle(\Delta x)^2\rangle k_\perp^2}. \quad (2.98)$$

This is the integral equation for the field line mean square deviation. This formula is correct for a Gaussian statistics and if the Corrsin approximation is valid. By applying the operator d/dz onto this result, we find

$$\frac{d}{dz}\left\langle (\Delta x(z))^2 \right\rangle = \frac{2}{B_0^2} \int d^3k \; P_{xx}(\mathbf{k}) \int_0^z dz' \; \cos\left(k_\parallel z'\right) e^{-\frac{1}{2}\langle(\Delta x)^2\rangle k_\perp^2}. \quad (2.99)$$

By considering the second derivative of the field line mean square deviation, an ordinary differential equation can be obtained

$$\frac{d^2}{dz^2}\left\langle (\Delta x(z))^2 \right\rangle = \frac{2}{B_0^2} \int d^3k \; P_{xx}(\mathbf{k}) \cos\left(k_\parallel z\right) \, e^{-\frac{1}{2}\langle(\Delta x)^2\rangle k_\perp^2}. \quad (2.100)$$

[1] For a diffusive behavior of field lines, the diffusion equation can be used and can be solved for sharp initial conditions. In this case, the assumption of a Gaussian statistics of the field lines is valid (see Chap. 1). For a nondiffusive field line behavior, as obtained for the slab/2D model, the assumption of a Gaussian statistics could be inaccurate. However, for cases where the field line behavior is close to the diffusive regime, (2.97) should be a good approximation.

This differential equation is very useful to describe field line random walk analytically.

As an example, we consider again the slab/2D composite model. In this case, (2.100) becomes

$$\frac{d^2}{dz^2}\left\langle(\Delta x(z))^2\right\rangle = \frac{2}{B_0^2}\int d^3k P_{xx}^{\text{slab}}(\mathbf{k})\cos(k_\parallel z) + \frac{2}{B_0^2}\int d^3k P_{xx}^{2D}(\mathbf{k}) e^{-\frac{1}{2}\langle(\Delta x)^2\rangle k_\perp^2}. \tag{2.101}$$

The first term can be expressed by the slab correlation function $R_{xx}^{\text{slab}}(\mathbf{k})$ (see (2.38)). For the 2D contribution, we can use (2.50) to find

$$\frac{d^2}{dz^2}\left\langle(\Delta x)^2\right\rangle = \frac{2}{B_0^2}R_{xx}^{\text{slab}}(z) + \frac{2\pi}{B_0^2}\int_0^\infty dk_\perp\, g^{2D}(k_\perp)\, e^{-\frac{1}{2}\langle(\Delta x)^2\rangle k_\perp^2}. \tag{2.102}$$

Now we consider the limit of $z \to \infty$ and, therefore, $\left\langle(\Delta x)^2\right\rangle \to \infty$. In this case, there is only a contribution to the integral when $k_\perp \to 0$. By assuming that the spectrum is constant in the energy range, we can approximate

$$\begin{aligned}\frac{d^2}{dz^2}\left\langle(\Delta x)^2\right\rangle &\approx \frac{2}{B_0^2}R_{xx}^{\text{slab}}(z\to\infty) + \frac{2\pi}{B_0^2}g^{2D}(0)\int_0^\infty dk_\perp\, e^{-\frac{1}{2}\langle(\Delta x)^2\rangle k_\perp^2}, \\ &\approx \frac{2\pi}{B_0^2}\sqrt{\frac{\pi}{2}}g^{2D}(0)\left[\left\langle(\Delta x)^2\right\rangle\right]^{-1/2}. \end{aligned} \tag{2.103}$$

Here we assumed an exponentially decaying slab correlation function (see (2.44)). Therefore, the slab contribution can be neglected in comparison to the 2D contribution. It can easily be demonstrated that, for $z \to \infty$, the solution of (2.103) has the form

$$\left\langle(\Delta x)^2\right\rangle = \left[\frac{9\pi}{2B_0^2}\sqrt{\frac{\pi}{2}}g^{2D}(0)\right]^{2/3} \mid z\mid^{4/3}. \tag{2.104}$$

The general solution contains two integration constants, which, in the limit $|z|\to\infty$, can be neglected. With (2.51) for the 2D wave spectrum, one finally finds

$$\left\langle(\Delta x)^2\right\rangle = \left[9C(\nu)\sqrt{\frac{\pi}{2}}l_{2D}\frac{\delta B_{2D}^2}{B_0^2}\right]^{2/3} \mid z\mid^{4/3}. \tag{2.105}$$

Thus, the running field line diffusion coefficient $d_{FL}(z)$ for slab/2D composite geometry has the form

$$d_{FL}(z) = \frac{1}{2}\frac{d}{dz}\left\langle(\Delta x)^2\right\rangle = \frac{2}{3}\left[9C(\nu)\sqrt{\frac{\pi}{2}}l_{2D}\frac{\delta B_{2D}^2}{B_0^2}\right]^{2/3} \mid z\mid^{1/3}, \tag{2.106}$$

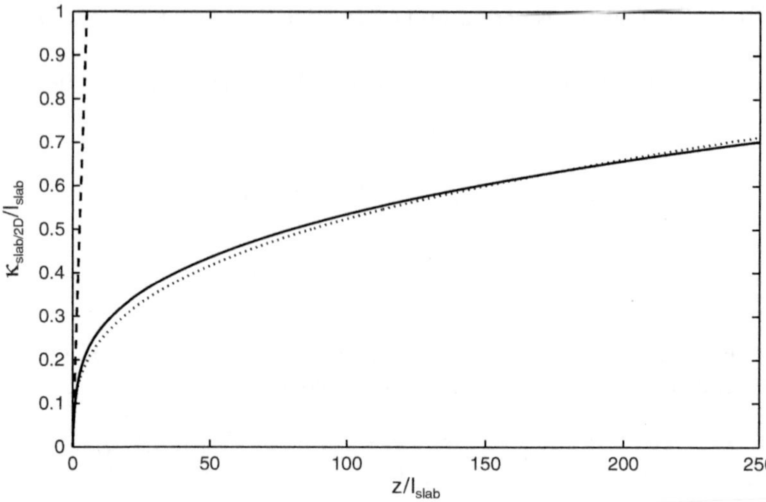

Fig. 2.6 The field line diffusion coefficient for the slab/2D composite model. Compared is the analytical result from (2.105) indicated by the *dotted line*, with the numerical result (*solid line*) obtained by solving (2.98) numerically. For completeness also the quasilinear result for the slab/2D model is shown (*dashed line*)

which is clearly a superdiffusive result. In Fig. 2.6 the field line diffusion coefficient

$$\kappa_{\mathrm{FL}}(z) = \frac{\left\langle (\Delta x)^2 \right\rangle}{2\mid z \mid} \tag{2.107}$$

is shown. Compared are the analytical result from (2.105) with the result obtained by solving (2.98) numerically. There, it is demonstrated that (2.105) is indeed the correct asymptotic behavior of field line random walk.

2.4.5 *The Diffusion Limit of Matthaeus et al*

The first authors who developed a nonlinear formulation for describing field line random walk analytically were Matthaeus et al. (1995). Here, we deduce the Matthaeus et al. approach as a special limit from (2.99). First, we assume that field line wandering behaves diffusively for all length scales. Then, one can use $\left\langle (\Delta x)^2 \right\rangle = 2\kappa_{\mathrm{FL}}|z|$ on the left- and right-hand side of (2.99) to get

$$\kappa_{\mathrm{FL}} = \frac{1}{B_0^2} \int \mathrm{d}^3k \; P_{xx}(\mathbf{k}) \int_0^\infty \mathrm{d}z \, \cos(k_\parallel z) \, \mathrm{e}^{-\kappa_{\mathrm{FL}} \, z \, k_\perp^2}. \tag{2.108}$$

Here, we used the field line diffusion coefficient κ_{FL}. The integral over z can be solved analytically and one gets

$$\kappa_{\mathrm{FL}} = \frac{1}{B_0^2} \int \mathrm{d}^3k \; P_{xx}(\mathbf{k}) \frac{\kappa_{\mathrm{FL}} k_\perp^2}{k_\parallel^2 + (\kappa_{\mathrm{FL}} k_\perp^2)^2}. \tag{2.109}$$

This formula is correct if field line random walk behaves diffusively and if the small length scales of the initial free-streaming regime are unimportant. For two-component turbulence, (2.109) can easily be evaluated and we find

$$\begin{aligned} \kappa_{\mathrm{FL}} &= \frac{\pi}{B_0^2} \int \mathrm{d}^3 \; P_{xx}^{\mathrm{slab}}(\mathbf{k}) \delta(k_\parallel) \\ &+ \frac{1}{B_0^2} \int \mathrm{d}^3 \; P_{xx}^{\mathrm{2D}}(\mathbf{k}) \frac{k_\perp^2}{\kappa_{\mathrm{FL}}}. \end{aligned} \tag{2.110}$$

With the slab diffusion coefficient κ_{slab} and with the parameter

$$\kappa_{\mathrm{2D}}^2 = \frac{1}{B_0^2} \int \mathrm{d}^3k \; k_\perp^{-2} P_{xx}^{\mathrm{2D}}(\mathbf{k}), \tag{2.111}$$

this becomes

$$\kappa_{\mathrm{FL}} = \kappa_{\mathrm{slab}} + \frac{\kappa_{\mathrm{2D}}^2}{\kappa_{\mathrm{FL}}}. \tag{2.112}$$

For the spectrum of (2.33), it can easily be demonstrated that the slab contribution behaves indeed diffusively (see (2.88)) and can, therefore, be represented by the slab diffusion coefficient κ_{slab}. Equation (2.112) is a quadratic formula and can easily be solved to get the Matthaeus et al. result

$$\kappa_{\mathrm{FL}} = \frac{\kappa_{\mathrm{slab}} + \sqrt{\kappa_{\mathrm{slab}}^2 + 4\kappa_{\mathrm{2D}}^2}}{2}. \tag{2.113}$$

For pure slab turbulence, we have $\kappa_{\mathrm{2D}} = 0$ and we get the expected result $\kappa = \kappa_{\mathrm{slab}}$. For pure 2D geometry, we have $\kappa_{\mathrm{slab}} = 0$ and we find $\kappa = \kappa_{\mathrm{2D}}$. Thus, the parameter κ_{2D} can be identified with the diffusion coefficient for pure 2D turbulence. However, for the standard spectrum of (2.51), we find

$$\kappa_{\mathrm{2D}}^2 = 2C(\nu) l_{\mathrm{2D}} \frac{\delta B_{\mathrm{2D}}^2}{B_0^2} \int_0^\infty \mathrm{d}k_\perp \; k_\perp^{-2} (1 + k_\perp^2 l_{\mathrm{2D}}^2)^{-\nu}, \tag{2.114}$$

which clearly does not converge ($\kappa_{\mathrm{2D}} \to \infty$). To enforce a finitely large parameter κ_{2D}, Matthaeus et al. introduced an additional lengthscale, namely the so-called ultra-scale $l_{2D,ultra}$. The ultra-scale is defined so, that the 2D diffusion coefficient has the form

$$\kappa_{2D} = l_{2D,ultra} \frac{\delta B_{2D}}{B_0}. \tag{2.115}$$

In this case an analytical form of the total field line diffusion coefficient can be given if the ultra-scale is known. However, as demonstrated in the previous paragraph the reason for $\kappa_{2D} \to \infty$ is the superdiffusive behavior. Thus, (2.113) cannot be confirmed by the more general calculations presented in Shalchi and Kourakis (2007). However, the formula from (2.109) can be applied in cases where field line wandering behaves indeed diffusively. Such cases can be obtained for certain forms of the 2D wave spectrum (see, e.g., Matthaeus et al. 2007).

2.5 Dynamical Turbulence and Plasma Wave Propagation Effects

So far, we have considered only magnetostatic turbulence where the correlation tensor is time-independent. In reality, however, the correlation tensors can also be a function of time:

$$\begin{aligned} R_{lm}(\mathbf{x},t) &= \langle \delta B_l(\mathbf{x},t)\delta B_m^*(0,0)\rangle, \\ P_{lm}(\mathbf{k},t) &= \langle \delta B_l(\mathbf{k},t)\delta B_m^*(\mathbf{k},0)\rangle. \end{aligned} \tag{2.116}$$

By assuming the same temporal behavior of all tensor components, we have, according to (2.8), $P_{lm}(\mathbf{k},t) = P_{lm}(\mathbf{k},0)\ \Gamma(\mathbf{k},t)$, with the dynamical correlation function $\Gamma(\mathbf{k},t)$. In the following, we discuss several models for $\Gamma(\mathbf{k},t)$.

2.5.1 *Damping and Random Sweeping Models*

One of the first authors discussing cosmic ray transport in dynamical turbulence were Bieber et al. (1994). In their article, the authors proposed two models for the dynamical correlation function:

$$\begin{aligned} \Gamma_{DT}(\mathbf{k},t) &= e^{-t/t_c} \quad \text{(damping model of dynamical turbulence)}, \\ \Gamma_{RS}(\mathbf{k},t) &= e^{-(t/t_c)^2} \quad \text{(random sweeping model)}, \end{aligned} \tag{2.117}$$

with the correlation time scale t_c. Bieber et al. (1994) estimated the correlation time as

$$t_c^{-1} = \alpha v_A |\mathbf{k}|. \tag{2.118}$$

Here, v_A is the Alfvén speed and α is a parameter that allows to adjust the strength of the dynamical effects, ranging from $\alpha = 0$ (magnetostatic turbulence) to $\alpha = 1$ (strongly dynamical turbulence). Bieber et al. (1994) also suggested that the parameter α could be interpreted as $\delta B/B_0$. In this case, the correlation time scale t_c becomes comparable to the eddy turnover time. Also, decorrelation effects related to plasma waves (see, e.g., Schlickeiser and Achatz (1993)) can be achieved by expressing α through parameters such as the plasma β (for a definition see also Schlickeiser and Achatz 1993). The damping model was originally introduced for a particle scattering study in dynamcial turbulence by Bieber and Matthaeus (1991, 1992). In this model, the dynamical correlation function has an exponential form, whereas in the random sweeping model $\Gamma(\mathbf{k}, t)$ has a Gaussian form.

2.5.2 *Plasma Wave Turbulence*

Another prominent model is the plasma wave model, which is discussed in Schlickeiser (2002). In this model, the dynamical correlation function has the form

$$\Gamma_{\mathrm{PW}}(\mathbf{k}, t) = \mathrm{e}^{\mathrm{i}\omega t - \gamma t}. \tag{2.119}$$

Here, ω is the plasma wave dispersion relation, and γ desribes plasma wave damping. Often, undamped plasma waves are considered, where $\Gamma_{\mathrm{PW}}(\mathbf{k}, t) = \mathrm{e}^{\mathrm{i}\omega t}$ and the dynamical correlation function is a purely oscillating function. Prominent examples for different plasma waves are Shear Alfvén waves, where $\omega = \pm v_A k_\parallel$, and fast magnetosonic waves, where $\omega = v_A k$.

At least within the plasma wave model also turbulent electric fields can be found. For undamped plasma waves, we can rewrite Faraday's law as

$$\nabla \times \mathbf{E} = -\frac{1}{c}\frac{\partial}{\partial t}\mathbf{B} \Rightarrow \mathbf{k} \times \mathbf{E} \sim \frac{\omega}{c}\mathbf{B}. \tag{2.120}$$

By assuming $\omega \sim v_A k$, we can estimate the strength of the electric fields by

$$\delta E \sim \frac{v_A}{c}\delta B \ll \delta B. \tag{2.121}$$

Therefore, plasma wave propagation effects as well as electric fields can be neglected for most test-particle studies, except in two cases:

1. Low energy cosmic rays, where the particle velocity becomes comparably small to the Alfvén speed ($v \approx v_A$)
2. Particle acceleration through the turbulent electric fields. In this case, a momentum diffusion coefficient can also be determined (see Sect. 1.6)

In Table 2.1, the different models for the dynamical correlation function are summarized. In the next paragraph, we consider a more advanced turbulence model that

Table 2.1 Previous models for the dynamical correlation function

Model	$\Gamma(\mathbf{k},t)$
Magnetostatic model	1
Damping model of dynamical turbulence	$e^{-\alpha v_A k t}$
Random sweeping model	$e^{-(\alpha v_A k t)^2}$
Plasma wave model for shear Alfvén waves	$e^{\pm i v_A k_\parallel t}$
Plasma wave model for fast mode waves	$e^{i v_A k t}$

Here, v_A is the Alfvén speed and α is a parameter that allows to adjust the strength of dynamical effects. Here we used $k = |\mathbf{k}|$

was proposed by Shalchi et al. (2006). This improved turbulence model takes into account dynamical as well as plasma wave effects.

2.5.3 The Nonlinear Anisotropic Dynamical Turbulence Model

Recently, an improved dynamical turbulence model, namely the *nonlinear anisotropic dynamical turbulence (NADT)* model, has been proposed by Shalchi et al. (2006). The NADT model was formulated for the slab/2D composite model, where, in general, we have the two different dynamical correlation functions $\Gamma^{\text{slab}}(k_\parallel, t)$ and $\Gamma^{2D}(k_\perp, t)$:

$$P_{lm}(\mathbf{k},t) = P_{lm}^{\text{slab}}(\mathbf{k})\,\Gamma^{\text{slab}}(k_\parallel,t) + P_{lm}^{2D}(\mathbf{k})\,\Gamma^{2D}(k_\perp,t). \tag{2.122}$$

In earlier treatments of dynamical turbulence, the decorrelation factors $\Gamma^i(\mathbf{k},t)$ were established by using simple approximations to the interactions responsible for the temporal decorrelation of excitations near the wave vector $\mathbf{k}$. In damping and random sweeping models, for example, a single parameter is introduced to estimate the rate of decorrelation at scale $1/k$, which is assumed to be related to the large scale Alfvén speed v_A (see, e.g., (2.118)). In the plasma wave model, the decorrelation rate is associated with oscillations at the Alfvén wave frequency. In recent years, however, there has been a more complete understanding of the time scales of magnetohydrodynamic (MHD) turbulence (e.g., Zhou et al. 2004) and of the relation they may have with the interactions between excitations that may be associated with either low frequency or wave-like components of the turbulence spectrum (Matthaeus et al. 1990; Tu and Marsch 1993; Oughton et al. 2006). In the context of the two component model, such ideas may be used to determine reasonable approximations to the functions $\Gamma^{\text{slab}}(\mathbf{k},t)$ and $\Gamma^{2D}(\mathbf{k},t)$ that appear in (2.122).

These functions are to be interpreted as the time decorrelation rates of the fluctuations near the wave vector $\mathbf{k}$, normalized to the associated energy, and are related to the ordinary two-time (Eulerian) correlation function $E(t) = \langle \delta B_l(\mathbf{x},t)\delta B_l^*(\mathbf{x},0)\rangle$ through $\int d^3k P_{ll}(\mathbf{k},t) = E(t)$. Clearly $\Gamma(\mathbf{k},0) \equiv 1$ while $\lim_{t\to\infty}\Gamma(\mathbf{k},t) = 0$. In various forms, these dynamical decorrelation functions occupy a central role in analytical turbulence closures (see, e.g., Zhou et al. 2004).

In establishing time scales for MHD turbulence, the distinction between those fluctuations having more wave-like dynamics and those having more hydrodynamic or zero frequency dynamics is crucial. For fluctuations that are predominantly incompressive (such as the solar wind), this characterization requires comparing estimates of the wave period with estimates of the nonlinear timescale or eddy turnover time. If the wave timescale is the shorter of the two, wave effects are significant even if there is turbulence. If the nonlinear time scale is shorter, nonlinear effects should dominate and wave effects are expected to be weaker.

In a practical sense, in which the spectrum is not made up of pure symmetry components, one would group with the slab population those wave vectors for which decorrelation occurs due to wave propagation, assumed here to be Alfvénic in character. This leads, in $\Gamma^{\rm slab}$, to a factor of an oscillatory nature, with the angular frequency given by $\omega = k_\parallel v_{\rm A}$. In addition, the slab-like component experiences resonant nonlinear triad interactions with the low frequency 2D component, as discussed by Shebalin (1983) and Oughton et al. (1994). Such interactions are characterized by a nonlinear time computed from the 2D turbulence, without further wave influence. For simplicity, we can estimate this influence by the *global* 2D nonlinear time scale, $\tau_{nl}^{\rm 2D} = \alpha^{-1} l_{\rm 2D}/Z$, where Z is the 2D turbulence amplitude (Oughton et al. 2006), α is a constant of order one related to the Karman–Taylor constants associated with the decay of the turbulence, and $l_{\rm 2D}$ is the spectral bendover scale of the 2D component.

For the 2D component, one reasonably groups together all the fluctuations for which wave-like propagation effects are weak. Therefore, in constructing $\Gamma^{\rm 2D}$, no oscillatory factor needs to appear. For $k_\perp$ sufficiently small, the 2D fluctuations lie in the energy range, and an estimate for the decorrelation rate is again the global rate $1/\tau_{nl}^{\rm 2D}$. For smaller 2D fluctuations, the decorrelation can be estimated using a steady inertial range $k_\perp^{-5/3}$ approximation, which is essentially what enters the steady Goldreich–Sridhar (1995) spectral theory. In this case, the nonlinear time scale is $\tau_{nl}^{\rm 2D}(k_\perp) = 1/(k_\perp u_{k_\perp})$, where the turbulence amplitude is $u_{k_\perp} = (k_\perp E(k_\perp))^{1/2}$ for 2D steady inertial range wave number spectrum $E(k_\perp) \sim k_\perp^{-5/3}$. This timescale is shorter than that of the energy range, and varies with $k_\perp$ according to $\tau_{nl}^{\rm 2D}(k_\perp) = \tau_{nl}^{\rm 2D}/(k_\perp l_{2D})^{2/3}$.

Assembling these results, we obtain within the NADT model

$$\Gamma^{\rm slab}(k_\parallel, t) = {\rm e}^{-\gamma^{\rm slab}\, t} \cdot {\rm e}^{{\rm i}\omega t}, \tag{2.123}$$

with

$$\gamma^{\rm slab} = (\tau_{nl}^{\rm 2D})^{-1} = \beta = \frac{Z}{l_{\rm 2D}} \tag{2.124}$$

and with the plasma wave dispersion relation of shear Alfvén waves

$$\omega = j v_{\mathrm{A}} k_{\parallel}. \tag{2.125}$$

For the 2D dynamical correlation function $\Gamma^{\mathrm{2D}}(\mathbf{k}, t)$, we have

$$\Gamma^{\mathrm{2D}}(k_{\perp}, t) = \mathrm{e}^{-\gamma^{\mathrm{2D}}\, t}, \tag{2.126}$$

with

$$\gamma^{\mathrm{2D}} = \beta \begin{cases} 1 & \text{for } k_{\perp} l_{\mathrm{2D}} \leq 1 \\ (k_{\perp} l_{\mathrm{2D}})^{2/3} & \text{for } k_{\perp} l_{\mathrm{2D}} \geq 1. \end{cases} \tag{2.127}$$

In (2.124), the parameter Z can be expressed by the strength of the 2D component $\delta B_{\mathrm{2D}}/B_0$ and the Alfvén Speed v_{A}:

$$Z = \sqrt{2}\frac{\delta B_{\mathrm{2D}}}{\sqrt{4\pi\rho_{\mathrm{d}}}} = \sqrt{2} v_{\mathrm{A}} \frac{\delta B_{\mathrm{2D}}}{B_0} \tag{2.128}$$

and we find

$$\beta = \sqrt{2}\frac{v_{\mathrm{A}}}{l_{\mathrm{2D}}}\frac{\delta B_{\mathrm{2D}}}{B_0}. \tag{2.129}$$

The parameter j in (2.125) is used to track the wave propagation direction ($j = +1$ is used for forward and $j = -1$ for backward to the ambient magnetic field propagating Alfvén waves). A lot of studies have addressed the direction of propagation of Alfvénic turbulence, see, for instance, Bavassano (2003). Most of the observations, which allow conclusions on space plasma and particle propagation properties, have been performed in the solar wind at 1 AU heliocentric distance. In this case, we can assume that all waves propagate forward, and we can therefore set $j = +1$. In general, one would expect that, closer to the Sun, most waves should propagate forward and, far away from the Sun, the wave intensities should be equal for both directions.

The model defined above takes into account the plasma wave character of the turbulence. In Chap. 8, the NADT model will be used to reproduce the observed cosmic ray parallel and perpendicular mean free paths in the heliosphere.

Chapter 3
The Quasilinear Theory

In this chapter, we discuss the interaction between charged particles and an astrophysical plasma. Historically, the first and most applied approach to determine spatial diffusion coefficients and other transport parameters is the so-called quasilinear theory (QLT). The quasilinear approximation is comparable to a first-order perturbation theory. In this chapter, different analytical and numerical results obtained within the magnetostatic slab, 2D, and composite model are presented. Hereafter we consider test-particle simulations and we compare their results with QLT. It is demonstrated that there are at least three cases for which QLT is insufficient. These three problems of QLT are the motivation for developing nonlinear transport theories.

3.1 The Quasilinear Approximation

During the previous decades, the quasilinear theory (QLT) of cosmic ray transport was the standard tool for calculating transport parameters. As demonstrated in Chap. 1, parallel and perpendicular mean free paths can be calculated using

$$\lambda_{\parallel} = \frac{3}{v}\kappa_{\parallel} = \frac{3v}{8}\int_{-1}^{+1} d\mu \frac{(1-\mu^2)^2}{D_{\mu\mu}(\mu)},$$
$$\lambda_{\perp} = \frac{3}{v}\kappa_{\perp} = \frac{3}{2v}\int_{-1}^{+1} d\mu \; D_{\perp}(\mu), \tag{3.1}$$

where the pitch-angle Fokker–Planck coefficient $D_{\mu\mu}$ and the Fokker–Planck coefficient of perpendicular diffusion $D_{\perp}$ have been used, which can be calculated by combining the TGK representation

$$D_{\mu\mu}(\mu) = \int_0^{\infty} dt \; \langle \dot{\mu}(t)\dot{\mu}(0)\rangle,$$
$$D_{ij}(\mu) = \int_0^{\infty} dt \; \langle \tilde{v}_i(t)\tilde{v}_j(0)\rangle, \quad i = x, y \tag{3.2}$$

A. Shalchi, *Nonlinear Cosmic Ray Diffusion Theories*,
Astrophysics and Space Science Library,
DOI: 10.1007/978-3-642-00309-7_3, © Springer-Verlag Berlin Heidelberg 2009

with the equations of motion

$$\dot{\mu} = \frac{1}{R_{\mathrm{L}}}\left(v_x \frac{\delta B_y(\mathbf{x})}{B_0} - v_y \frac{\delta B_x(\mathbf{x})}{B_0}\right), \tag{3.3}$$

$$\tilde{v}_i = v_z \frac{\delta B_i(\mathbf{x})}{B_0}, \quad i = x, y. \tag{3.4}$$

Here, we assume purely magnetic fluctuations ($\delta E_i = 0$). Furthermore, the latter equation, which describes guiding centers that follow magnetic field lines, is valid only if the perturbation terms ($v_i \delta B_j$) are varying slowly within an unperturbed gyroperiod (see Chap. 1).

Within the quasilinear approximation, we replace the particle velocity, v_i, and the position of the particle, x_i, on the right hand side of (3.3) and (3.4) by the unperturbed trajectory (the particle orbit for $\delta B_i = 0$)

$$x_i(t), v_i(t) \quad \stackrel{\text{QLT}}{\Longrightarrow} \quad x_{i,0}(t), v_{i,0}(t). \tag{3.5}$$

Immediately the question arises how this approximation can be justified. Within the TGK formulation, a diffusion coefficient is calculated by applying the operator $\int_0^\infty dt$ onto the perturbation terms. Therefore, the particle orbit as well as the particle velocity have to be known for all time-scales. Within QLT, we replace the true particle orbit by the unperturbed trajectory. Because of the permanent interaction with the turbulent fields, the deviation between the unperturbed orbit and the true motion becomes larger during time. Therefore, for late times, the quasilinear approximation becomes less and less valid. Hence, QLT is correct only if later times are unimportant. As demonstrated in Sect. 3.7 and in Chap. 7, the application of QLT leads to several problems. For instance, it is demonstrated systematically in Chap. 7 that for perpendicular transport larger time-scales are important. Therefore, it might be more reasonable to replace the particle motion by a diffusive motion. This approach is one of the basic assumptions that enter the nonlinear guiding center theory (NLGC theory) discussed in Chap. 4. Some authors (e.g., Schlickeiser 2002) justified QLT by assuming smallness of the turbulent fields in comparison to the mean magnetic field ($\delta B_i \ll B_0$). However, as demonstrated in Sect. 3.7, the failure of QLT is not always related to the strength of the turbulent fields. For instance, QLT is always invalid for perpendicular transport, for pitch-angle diffusion at 90° ($\mu = 0$), and for particle transport in 2D-like turbulence.

On the other hand, it should be noted that for some cases (parallel transport in slab turbulence with a Kolmogorov wave spectrum) QLT works very well. Furthermore, if QLT is combined with improved turbulence models (e.g., the NADT model discussed in Chap. 2), we can achieve agreement with measurements of parallel mean free paths in the heliosphere (see Chap. 8). Another advantage of QLT is the systematic character of the theory. Nonlinear theories are mostly based on certain assumptions and models. Therefore, it is an interesting matter to combine

QLT with different turbulence models and to compare the results with test-particle simulations (see end of this chapter) and observations (see Chap. 8).

With the unperturbed trajectory of (1.11), the unperturbed velocity of (1.10), and by using (3.3) in combination with the TGK formulation (3.2), it is straightforward to calculate quasilinear Fokker–Planck coefficients. In the following, we investigate the pitch-angle Fokker–Planck coefficient, the Fokker–Planck coefficient of perpendicular diffusion, and parallel as well as perpendicular mean free paths for pure slab, pure 2D, and composite turbulence within the magnetostatic limit.

3.2 General Forms of Quasilinear Fokker–Planck Coefficients

Before we consider quasilinear transport in certain turbulence models, we derive general forms of the pitch-angle Fokker–Planck coefficient, following the analysis of Teufel and Schlickeiser (2002), and the Fokker–Planck coefficient of perpendicular diffusion.

3.2.1 General Form of the Pitch-angle Fokker–Planck Coefficient

The starting point of our calculations is the parallel component of the equation of motion (see (3.3)). With the application of QLT (3.5) and the unperturbed orbit ((1.10) with (1.11)), we find

$$\dot{\mu} = \frac{\mathrm{d}\mu}{\mathrm{d}t} = \frac{\Omega\sqrt{1-\mu^2}}{B_0}\left[\delta B_y(\mathbf{x}_0)\,\cos\Phi(t) - \delta B_x(\mathbf{x}_0)\,\sin\Phi(t)\right], \tag{3.6}$$

with $\Phi(t) = \Phi_0 - \Omega t$. By using

$$\begin{aligned} \cos\Phi &= \frac{1}{2}\left(\mathrm{e}^{\mathrm{i}\Phi} + \mathrm{e}^{-\mathrm{i}\Phi}\right), \\ \sin\Phi &= \frac{1}{2\mathrm{i}}\left(\mathrm{e}^{\mathrm{i}\Phi} - \mathrm{e}^{-\mathrm{i}\Phi}\right) \end{aligned} \tag{3.7}$$

and by introducing left-handed and right-handed components for the turbulent fields

$$\begin{aligned} \delta B_{\mathrm{L}} &= \frac{1}{\sqrt{2}}\left(\delta B_x + \mathrm{i}\delta B_y\right), \\ \delta B_{\mathrm{R}} &= \frac{1}{\sqrt{2}}\left(\delta B_x - \mathrm{i}\delta B_y\right), \end{aligned} \tag{3.8}$$

this can be written as

$$\dot{\mu} = \frac{\mathrm{i}\Omega}{\sqrt{2}B_0}\sqrt{1-\mu^2}\left[\delta B_{\mathrm{R}}(\mathbf{x}_0(t))\,\mathrm{e}^{\mathrm{i}\Phi(t)} - \delta B_{\mathrm{L}}(\mathbf{x}_0(t))\,\mathrm{e}^{-\mathrm{i}\Phi(t)}\right]. \tag{3.9}$$

Thus, the pitch-angle Fokker–Planck coefficient of (3.2) has the form

$$\begin{aligned} D_{\mu\mu} =& \frac{\Omega^2(1-\mu^2)}{2B_0^2}\Re\mathfrak{e}\int_0^\infty \mathrm{d}t\left[R_{\mathrm{RR}}\,\mathrm{e}^{-\mathrm{i}\Omega t}\right. \\ & \left. - R_{\mathrm{RL}}\,\mathrm{e}^{+2\mathrm{i}\Phi_0-\mathrm{i}\Omega t} - R_{\mathrm{LR}}\,\mathrm{e}^{-2\mathrm{i}\Phi_0+\mathrm{i}\Omega t} + R_{\mathrm{LL}}\,\mathrm{e}^{+\mathrm{i}\Omega t}\right], \end{aligned} \tag{3.10}$$

with

$$R_{lm}(\mathbf{x}_0(t),t,\mathbf{x}_0(0),0) = \langle \delta B_l(\mathbf{x}_0(t),t)\delta B_m^*(\mathbf{x}_0(0),0)\rangle. \tag{3.11}$$

Now we use the Fourier transforms of the turbulent magnetic fields

$$\delta B_{\mathrm{L,R}}(\mathbf{x}_0,t) = \int \mathrm{d}^3k\;\delta B_{\mathrm{L,R}}(\mathbf{k},t)\,\mathrm{e}^{\mathrm{i}\mathbf{k}\cdot\mathbf{x}_0} \tag{3.12}$$

to obtain

$$R_{lm}(\mathbf{x}_0(t),t,\mathbf{x}_0(0),0) = \int \mathrm{d}^3k \int \mathrm{d}^3k' \left\langle \delta B_l(\mathbf{k},t)\delta B_m^*(\mathbf{k}',0)\right\rangle \mathrm{e}^{\mathrm{i}\left[\mathbf{k}\cdot\mathbf{x}_0(t)-\mathbf{k}'\cdot\mathbf{x}_0(0)\right]}. \tag{3.13}$$

With spherical coordinates for the wave vector

$$\begin{aligned} k_x &= k_\perp \cos\Psi, \\ k_y &= k_\perp \sin\Psi, \\ k_z &= k_\parallel, \end{aligned} \tag{3.14}$$

we derive

$$\mathbf{k}\cdot\mathbf{x}_0(t) = \mathbf{k}\cdot\mathbf{x}_0(0) + \frac{k_\perp v_\perp}{\Omega}\left[\sin(\Psi-\Phi_0+\Omega t) - \sin(\Psi-\Phi_0)\right] + k_\parallel v_\parallel t. \tag{3.15}$$

Therefore, it can easily be demonstrated that

$$R_{lm}(\mathbf{x}_0(t),t,\mathbf{x}_0(0),0) = \int \mathrm{d}^3k\;P_{lm}(\mathbf{k},t)\,\mathrm{e}^{\mathrm{i}W[\sin(\Psi-\Phi_0+\Omega t)-\sin(\Psi-\Phi_0)]+\mathrm{i}v_\parallel k_\parallel}, \tag{3.16}$$

where we used (2.4) and

$$W = \frac{k_\perp v_\perp}{\Omega} = k_\perp R_{\mathrm{L}}\sqrt{1-\mu^2}. \tag{3.17}$$

With (2.55), we obtain

$$R_{ij}(\mathbf{x}_0(t), t, \mathbf{x}_0(0), 0) = \int \mathrm{d}^3k \; P_{ij}(\mathbf{k}, t) \sum_{n,m=-\infty}^{\infty} J_n(W) J_m(W) \times \mathrm{e}^{\mathrm{i}n(\Psi - \Phi_0 + \Omega t) - \mathrm{i}m(\Psi - \Phi_0) + \mathrm{i}v_\parallel k_\parallel}. \quad (3.18)$$

Equation (3.18) can now be combined with (3.10) to yield

$$\begin{aligned} D_{\mu\mu} = {} & \frac{\Omega^2(1-\mu^2)}{2B_0^2} \Re\mathfrak{e} \int_0^\infty \mathrm{d}t \int \mathrm{d}^3k \sum_{n,m=-\infty}^{\infty} J_n(W) J_m(W) \\ & \times \mathrm{e}^{\mathrm{i}n(\Psi - \Phi_0 + \Omega t) - \mathrm{i}m(\Psi - \Phi_0) + \mathrm{i}v_\parallel k_\parallel} \\ & \times \left[P_{\mathrm{RR}}(\mathbf{k}, t)\, \mathrm{e}^{-\mathrm{i}\Omega t} - P_{\mathrm{RL}}(\mathbf{k}, t)\, \mathrm{e}^{+2\mathrm{i}\Phi_0 - \mathrm{i}\Omega t} \right. \\ & \left. - P_{\mathrm{LR}}(\mathbf{k}, t)\, \mathrm{e}^{-2\mathrm{i}\Phi_0 + \mathrm{i}\Omega t} + P_{\mathrm{LL}}(\mathbf{k}, t)\, \mathrm{e}^{+\mathrm{i}\Omega t} \right]. \end{aligned} \quad (3.19)$$

Because of the assumption of axisymmetric turbulence, we can apply the operator

$$\frac{1}{2\pi} \int_0^{2\pi} \mathrm{d}\Phi_0 \quad (3.20)$$

on (3.19) to obtain

$$\begin{aligned} D_{\mu\mu} = {} & \frac{\Omega^2(1-\mu^2)}{2B_0^2} \Re\mathfrak{e} \sum_{n=-\infty}^{\infty} \int_0^\infty \mathrm{d}t \int \mathrm{d}^3k \, \mathrm{e}^{\mathrm{i}(k_\parallel v_\parallel + n\Omega)t} \\ & \times \left[J_{n+1}^2(W)\, P_{\mathrm{RR}}(\mathbf{k}, t) + J_{n-1}^2(W)\, P_{\mathrm{LL}}(\mathbf{k}, t) \right. \\ & \left. - J_{n+1}(W)\, J_{n-1}(W) \left(P_{\mathrm{RL}}(\mathbf{k}, t)\, \mathrm{e}^{+2\mathrm{i}\Psi} + P_{\mathrm{LR}}(\mathbf{k}, t)\, \mathrm{e}^{-2\mathrm{i}\Psi} \right) \right]. \end{aligned} \quad (3.21)$$

To perform the time integration, we have to specify the time dependence of the tensor P_{lm}. By assuming the form given by (2.8), we find

$$\begin{aligned} D_{\mu\mu} = {} & \frac{\Omega^2(1-\mu^2)}{2B_0^2} \sum_{n=-\infty}^{\infty} \int \mathrm{d}^3k \left[J_{n+1}^2\left(\frac{k_\perp v_\perp}{\Omega}\right) P_{\mathrm{RR}}(\mathbf{k}) + J_{n-1}^2\left(\frac{k_\perp v_\perp}{\Omega}\right) \right. \\ & \left. P_{\mathrm{LL}}(\mathbf{k}) - J_{n+1}\left(\frac{k_\perp v_\perp}{\Omega}\right) J_{n-1}\left(\frac{k_\perp v_\perp}{\Omega}\right) \left(P_{\mathrm{RL}}(\mathbf{k})\, \mathrm{e}^{+2\mathrm{i}\Psi} + P_{\mathrm{LR}}(\mathbf{k})\, \mathrm{e}^{-2\mathrm{i}\Psi} \right) \right] R_n(\mathbf{k}), \end{aligned} \quad (3.22)$$

with the quasilinear resonance function

$$R_n(\mathbf{k}) = \Re\mathfrak{e} \int_0^\infty \mathrm{d}t \, \mathrm{e}^{\mathrm{i}(k_\parallel v_\parallel + n\Omega)t} \, \Gamma(\mathbf{k}, t). \quad (3.23)$$

For magnetostatic turbulence, we have $\Gamma(\mathbf{k}, t) = 1$ and thus

$$R_n^{\mathrm{MS}}(\mathbf{k}) = \Re\mathfrak{e} \int_0^\infty \mathrm{d}t\, \mathrm{e}^{\mathrm{i}(k_\parallel v\mu + n\Omega)t} = \pi\delta(k_\parallel v\mu + n\Omega). \tag{3.24}$$

Obviously, within magnetostatic QLT, we find a sharp (gyro-)resonance function. The gyroresonance condition is

$$\frac{v_\parallel}{\Omega} = k_\parallel^{-1}. \tag{3.25}$$

The left-hand side of (3.25) is the parallel distance that the particle travels within one unperturbed gyroperiod Ω^{-1}. There is only interaction if this distance is equal to the wavelength of the turbulence $k_\parallel^{-1}$. It should be emphasized that the reason for the sharp resonance condition is the assumption of unperturbed orbits. In reality, however, the particle permanently experiences scattering, thus the true particle orbit becomes more and more uncertain during time. Obviously, the real resonance function is broadened. The accurate determination of the true resonance function is the basis of a nonlinear transport theory (see Chaps. 4–7).

3.2.2 *General Form of the Fokker–Planck Coefficient of Perpendicular Diffusion*

Here, we start with the assumption that guiding centers follow magnetic field lines (3.4),

$$\tilde{v}_i = v_z \frac{\delta B_i(\mathbf{x})}{B_0}, \tag{3.26}$$

for purely magnetic fluctuations and $\delta B_z = 0$. These formulas can easily be combined with QLT and the TGK formulation (3.2) to find

$$D_{ij} = \left(\frac{v\mu}{B_0}\right)^2 \int_0^\infty \mathrm{d}t\, R_{ij}(t). \tag{3.27}$$

For the magnetic correlation function $R_{ij}(t)$, we can use (3.16). By applying the initial gyrophase averaging operator of (3.20) onto this formula, we find

$$R_{ij}(t) = \int \mathrm{d}^3k\; P_{ij}(\mathbf{k}, t) \sum_{n=-\infty}^{\infty} J_n^2(W) e^{i(v_\parallel k_\parallel + n\Omega)}, \tag{3.28}$$

which can be combined with (3.27). By assuming the form of (2.8) for P_{ij}, we obtain

$$D_{ij} = \left(\frac{v\mu}{B_0}\right)^2 \int \mathrm{d}^3k\; P_{ij}(\mathbf{k}) \sum_{n=-\infty}^{\infty} J_n^2(W) R_n(\mathbf{k}), \tag{3.29}$$

where we again used the resonance function $R_n(\mathbf{k})$ of (3.23).

3.3 Standard QLT (Magnetostatic Slab Turbulence)

Here, we combine QLT with a magnetostatic slab model (see Chap. 2), often referred to as Standard QLT. The combination of QLT and magnetostatic turbulence has been used in the original Jokipii (1966) paper, where, for the first time, the QLT formulation was applied onto cosmic ray transport. By using the wave spectrum of (2.33), we determine the transport parameters $D_{\mu\mu}$, $D_\perp$, $\lambda_\parallel$, and $\lambda_\perp$.

3.3.1 The Pitch-angle Fokker–Planck Coefficient

A possible approach to determine $D_{\mu\mu}^{\text{slab}}$ would be the application of (3.22). For the slab model, however, it is more convenient to start directly with the equations of motion. By combining (3.3) with (3.2), we find within QLT

$$D_{\mu\mu} = \frac{1}{B_0^2 R_\text{L}^2} \Re\mathfrak{e} \int_0^\infty dt \, \left[v_x(t) v_x(0) R_{yy}(t) + v_y(t) v_y(0) R_{xx}(t) \right], \tag{3.30}$$

with the correlation functions from (2.1). Note that, in the slab model, we have $R_{xy} = R_{yx} = 0$. With (2.7) and (2.32), we find

$$R_{ii}(t) = \int d^3k \; P_{ii}(\mathbf{k}) \, e^{i\mathbf{kx}} = 4\pi \int_0^\infty dk_\parallel \; g^{\text{slab}}(k_\parallel) \, e^{ik_\parallel z}. \tag{3.31}$$

Furthermore, we can set the initial gyrophase to zero due to the axisymmetric assumption ($\Phi_0 = 0$):

$$\begin{aligned} v_x(t) &= v_\perp \cos(\Omega t), & v_x(0) &= v_\perp, \\ v_y(t) &= -v_\perp \sin(\Omega t), & v_y(0) &= 0. \end{aligned} \tag{3.32}$$

Thus, we obtain

$$v_x(t) v_x(0) R_{yy}(t) + v_y(t) v_y(0) R_{xx}(t) = \frac{v^2(1-\mu^2)}{2} \left(e^{+i\Omega t} + e^{-i\Omega t} \right) R_{yy}(t). \tag{3.33}$$

With (3.30), we find

$$D_{\mu\mu} = \frac{2\pi v^2 (1-\mu^2)}{B_0^2 R_L^2} \int_0^\infty dk_\parallel \; g^{\text{slab}}(k_\parallel) \sum_{n=\pm 1} R_n^{\text{slab}}(k_\parallel), \tag{3.34}$$

where we used the quasilinear resonance function from (3.24), which, for the slab model, is given by

$$R_\pm^{\text{slab}}(k_\parallel) = \pi \delta(k_\parallel v \mu \pm \Omega). \tag{3.35}$$

Because of the sharp resonance function, (3.34) becomes

$$D_{\mu\mu} = \frac{2\pi^2\Omega^2(1-\mu^2)}{B_0^2 v\mu} g^{\text{slab}}\left(k_\parallel = \frac{\Omega}{v\mu}\right). \tag{3.36}$$

With the spectrum from (2.33), we find

$$D_{\mu\mu} = \frac{\pi C(\nu)v}{l_{\text{slab}}}\frac{\delta B_{\text{slab}}^2}{B_0^2}(1-\mu^2)\mu^{2\nu-1}R^{2\nu-2}(1+\mu^2R^2)^{-\nu}, \tag{3.37}$$

where the dimensionless rigiditiy $R = R_L/l_{\text{slab}} = v/(\Omega l_{\text{slab}})$ was introduced.

3.3.2 The Parallel Mean Free Path

By using (3.2), we find for the parallel mean free path

$$\lambda_\parallel = \frac{3l_{\text{slab}}}{4\pi C(\nu)}\left(\frac{B_0}{\delta B_{\text{slab}}}\right)^2 R^{2-2\nu}\int_0^1 \mathrm{d}\mu(1-\mu^2)\mu^{1-2\nu}(1+\mu^2R^2)^{\nu}, \tag{3.38}$$

which can be expressed in terms of hypergeometric functions (see Gradshteyn and Ryzhik 2000, (3.194.1)):

$$\lambda_\parallel = \frac{3l_{\text{slab}}}{8\pi C(\nu)}\left(\frac{B_0}{\delta B_{\text{slab}}}\right)^2 R^{2-2\nu}\left[\frac{1}{1-\nu}\,{}_2F_1\left(1-\nu,-\nu,2-\nu,-R^2\right)\right.$$
$$\left.-\frac{1}{2-\nu}\,{}_2F_1\left(2-\nu,-\nu,3-\nu,-R^2\right)\right]. \tag{3.39}$$

For small rigidities ($R \ll 1$), this result can be further simplified by using the approximation (Abramowitz and Stegun 1974):

$$ {}_2F_1\,(a,b,c,0) \approx 1 \tag{3.40}$$

and, in this limit, we find

$$\lambda_\parallel(R \ll 1) = \frac{3l_{\text{slab}}}{8\pi C(\nu)}\left(\frac{B_0}{\delta B_{\text{slab}}}\right)^2 R^{2-2\nu}\left[\frac{1}{1-\nu}-\frac{1}{2-\nu}\right] \tag{3.41}$$

and, therefore, $\lambda_\parallel(R \ll 1) \sim R^{2-2\nu}$. For a Kolmogorov (1941) spectrum ($2\nu = 5/3$), we thus have $\lambda_\parallel \sim R^{1/3}$. In the limit of large rigidities, $R \gg 1$, we can also approximate the hypergeometric function (see e.g., Abramowitz and Stegun 1974) to obtain

$$\lambda_\parallel(R \gg 1) = \frac{3l_{\text{slab}}}{16\pi C(\nu)}\left(\frac{B_0}{\delta B_{\text{slab}}}\right)^2 R^2. \tag{3.42}$$

To obtain one simple formula for the parallel mean free path, we can simply add (3.41) and (3.42) to find

$$\lambda_{\parallel} \approx \frac{3 l_{\rm slab}}{16\pi C(\nu)} \left(\frac{B_0}{\delta B_{\rm slab}} \right)^2 R^{2-2\nu} \left[\frac{2}{(1-\nu)(2-\nu)} + R^{2\nu} \right]. \tag{3.43}$$

By considering the two limits $R \ll 1$ and $R \gg 1$, we can easily recover (3.41) and (3.42). In Sect. 3.4, we will compare these results with test-particle simulations to explore the validity of QLT.

3.3.3 The Perpendicular Mean Free Path

In the same manner, we can also obtain analytical expressions for the perpendicular mean free path. Here, we combine (3.4) with QLT:

$$v_x = v_z \frac{\delta B_x(z)}{B_0} \overset{\rm QLT}{=} v\mu \frac{\delta B_x(z_0(t))}{B_0}. \tag{3.44}$$

By applying (3.2), we find

$$D_{xx} = \int_0^\infty {\rm d}t \, \langle v_x(t) v_x(0) \rangle = \frac{v^2 \mu^2}{B_0^2} \int_0^\infty {\rm d}t \, \langle \delta B_x(z) \delta B_x^*(0) \rangle. \tag{3.45}$$

In QLT, we have $z_0(t) = v\mu t$ and thus

$$D_{xx} = \frac{v|\mu|}{B_0^2} \int_0^\infty {\rm d}z \, \langle \delta B_x(z) \delta B_x^*(0) \rangle. \tag{3.46}$$

The integral in (3.46) is proportional to the correlation length (2.45), which can be expressed by the slab bendover scale (2.48):

$$D_{xx} = \frac{v|\mu|}{2 B_0^2} \delta B_{\rm slab}^2 l_{\rm c} = \pi C(\nu) l_{\rm slab} v |\mu| \frac{\delta B_{\rm slab}^2}{B_0^2}. \tag{3.47}$$

For axisymmetric turbulence, we have $D_{\perp} = D_{xx} = D_{yy}$. Thus, we finally find for the perpendicular mean free path

$$\lambda_{\perp} = \frac{3}{v} \kappa_{\perp} = \frac{3}{2v} \int_{-1}^{+1} {\rm d}\mu \, D_{xx} = \frac{3\pi}{2} C(\nu) l_{\rm slab} \frac{\delta B_{\rm slab}^2}{B_0^2} =: \lambda_{\perp,\rm FLRW}. \tag{3.48}$$

This result is often referred to as the field line random walk limit (FLRW limit, Jokipii 1966). In this case, the perpendicular diffusion coefficient of the particle $\kappa_{\perp}$

can be expressed by the field line diffusion coefficient κ_{FL} (see Sect. 2.4)

$$\kappa_{\perp} = \frac{v}{2}\kappa_{\mathrm{FL}}. \tag{3.49}$$

Within QLT, we assume that the particle moves with constant parallel velocity along the field lines. In reality, however, the particle experiences parallel scattering while it moves through the turbulence. Thus, perpendicular transport and parallel transport occur simultaneously (compound diffusion, see Chap. 7).

3.4 Quasilinear Theory for Magnetostatic 2D Turbulence

In Chap. 2, we have seen that solar wind turbulence comprises a strong variation of turbulent magnetic fields across the mean magnetic field. Therefore, it is important to study the interaction between charged test-particles and 2D modes. To investigate pitch-angle transport in the 2D model, we discuss two different approaches.

3.4.1 *Pitch-angle Diffusion in Pure 2D Turbulence by Using the Traditional Approach*

Here, we follow the analysis of Shalchi and Schlickeiser (2004a) to determine the pitch-angle Fokker–Planck coefficient for pure 2D. The starting point is (3.22), which is also valid for pure 2D. In case of 2D turbulence, the tensor $P_{lm}(\mathbf{k})$ is given by (2.50) so that we obtain

$$\begin{aligned}
P_{\mathrm{LL}}(\mathbf{k}) &= P_{\mathrm{RR}}(\mathbf{k}) = \frac{1}{2} g^{2\mathrm{D}}(k_{\perp}) \frac{\delta(k_{\parallel})}{k_{\perp}}, \\
P_{\mathrm{LR}}(\mathbf{k}) &= \frac{1}{2} g^{2\mathrm{D}}(k_{\perp}) \frac{\delta(k_{\parallel})}{k_{\perp}} \left[-\cos(2\Psi) - \mathrm{i}\,\sin(2\Psi)\right], \\
P_{\mathrm{RL}}(\mathbf{k}) &= \frac{1}{2} g^{2\mathrm{D}}(k_{\perp}) \frac{\delta(k_{\parallel})}{k_{\perp}} \left[-\cos(2\Psi) + \mathrm{i}\,\sin(2\Psi)\right]
\end{aligned} \tag{3.50}$$

for the different helical components. To proceed, we express the trigonometric functions in (3.50) by exponential functions and use Bessel function identities (see e.g., Abramowitz and Stegun 1974),

$$\begin{aligned}
& J_{n+1}^2(W)\, P_{\mathrm{RR}}(\mathbf{k}) + J_{n-1}^2(W)\, P_{\mathrm{LL}}(\mathbf{k}) \\
& - J_{n+1}(W)\, J_{n-1}(W) \left(P_{\mathrm{RL}}(\mathbf{k})\, \mathrm{e}^{+2\mathrm{i}\Psi} + P_{\mathrm{LR}}(\mathbf{k})\, \mathrm{e}^{-2\mathrm{i}\Psi}\right) \\
& = \frac{1}{2} g^{2\mathrm{D}}(k_{\perp}) \frac{\delta(k_{\parallel})}{k_{\perp}} \left[J_{n+1}^2(W) + J_{n-1}^2(W) + 2 J_{n+1}(W) J_{n-1}(W)\right],
\end{aligned}$$

$$= \frac{1}{2} g^{2D}(k_\perp) \frac{\delta(k_\parallel)}{k_\perp} \left[J_{n+1}(W) + J_{n-1}(W)\right]^2,$$
$$= 2g^{2D}(k_\perp) \frac{\delta(k_\parallel)}{k_\perp} \frac{n^2 J_n^2(W)}{W^2}, \tag{3.51}$$

to find

$$D_{\mu\mu}^{2D} = \frac{\Omega^2(1-\mu^2)}{B_0^2} \int d^3k\, g^{2D}(k_\perp) \times \sum_{n=-\infty}^{\infty} R_n(\mathbf{k}) \frac{\delta(k_\parallel)}{k_\perp} \frac{n^2 J_n^2(W)}{W^2},$$
$$= \frac{2\pi\Omega^2(1-\mu^2)}{B_0^2} \int_0^\infty dk_\perp\, g^{2D}(k_\perp) \times \sum_{n=-\infty}^{\infty} R_n(k_\perp, k_\parallel = 0) \frac{n^2 J_n^2(W)}{W^2}. \tag{3.52}$$

Obviously, there is no contribution for $n = 0$ and thus

$$D_{\mu\mu}^{2D} = \frac{4\pi\Omega^2(1-\mu^2)}{B_0^2} \int_0^\infty dk_\perp\, g^{2D}(k_\perp) \sum_{n=1}^{\infty} R_n(k_\perp, k_\parallel = 0) \frac{n^2 J_n^2(W)}{W^2}. \tag{3.53}$$

This is the general form of the pitch-angle Fokker–Planck coefficient for pure 2D turbulence. This formula can also be applied within the weakly nonlinear approach (see Chap. 5) and for dynamical turbulence (see Chap. 8). For magnetostatic turbulence, the resonance function is $R_n(k_\perp, k_\parallel = 0) = \pi\delta(n\Omega) = 0$ for $n \neq 0$. Thus, we find vanishing pitch-angle scattering for pure 2D ($D_{\mu\mu}^{2D} = 0$) and QLT.

3.4.2 *Pitch-angle Diffusion in Pure 2D Turbulence by Using a Vector-potential Approach*

In the previous paragraph, it was already demonstrated that, for magnetostatic turbulence, we have $D_{\mu\mu}^{2D} = 0$. By using a vector potential approach, a different proof of this behavior is presented here. By using the magnetostatic vector potential $\delta A_z(x, y)$ (see Sect. 2.3), we can rewrite the parallel component of the equation of motion (3.3) as

$$\dot{\mu} = -\frac{\Omega}{vB_0}\left(v_x \partial_x \delta A_z + v_y \partial_y \delta A_z\right) = -\frac{\Omega}{vB_0} \frac{d}{dt} \delta A_z(x, y) \tag{3.54}$$

and, therefore, we obtain a direct relation between the change of the vector potential due to the particle motion and the change of the pitch-angle cosine:

$$\Delta\mu = -\frac{\Omega}{vB_0}\Delta\delta A_z, \tag{3.55}$$

where we used $\Delta\mu = \mu(t) - \mu(0)$ and $\Delta\delta A_z = \delta A_z(t) - \delta A_z(0)$. By using a Fourier representation, we obtain

$$\Delta\mu = -\frac{\Omega}{vB_0}\int \mathrm{d}^3k\; \delta A_z(\mathbf{k})\left[\mathrm{e}^{\mathrm{i}\mathbf{k}\cdot\mathbf{x}(t)} - \mathrm{e}^{\mathrm{i}\mathbf{k}\cdot\mathbf{x}(0)}\right], \tag{3.56}$$

which is an exact result for magnetostatic 2D turbulence. To proceed, we apply again QLT (3.5) and assume homogeneous turbulence. For the mean square displacement of the pitch-angle, we then find

$$\langle\Delta\mu\Delta\mu^*\rangle = 2\left(\frac{\Omega}{vB_0}\right)^2\int \mathrm{d}^3k\; \left\langle\delta A_z(\mathbf{k})\delta A_z^*(\mathbf{k})\right\rangle\left[1 - \cos\left(\mathbf{k}\Delta\mathbf{x}_0\right)\right], \tag{3.57}$$

where we used $\Delta\mathbf{x}_0 = \mathbf{x}_0(t) - \mathbf{x}_0(0)$ for the unperturbed orbit. Next, we define the time-dependent pitch-angle Fokker–Planck coefficient as

$$D_{\mu\mu}(t) := \frac{\left\langle(\Delta\mu)^2\right\rangle}{2t} \equiv \frac{\langle\Delta\mu\Delta\mu^*\rangle}{2t}, \tag{3.58}$$

which can easily be combined with (3.57) to obtain

$$D_{\mu\mu}(t) = \frac{1}{2t}\left(\frac{\Omega}{vB_0}\right)^2\int \mathrm{d}^3k\; A_{zz}(\mathbf{k})\left[2 - 2\cos\left(\mathbf{k}\Delta\mathbf{x}_0\right)\right]. \tag{3.59}$$

For the correlation function of the vector potential, A_{zz}, we can use (2.70), and for the 2D wave spectrum, we use (2.51). Hence, we can write

$$\begin{aligned}
D_{\mu\mu}(t) &= \frac{2C(\nu)l_{2\mathrm{D}}\delta B_{2\mathrm{D}}^2}{\pi t}\left(\frac{\Omega}{vB_0}\right)^2\int \mathrm{d}^2k\; \frac{1}{k_\perp^3}\left(1 + k_\perp^2 l_{2\mathrm{D}}^2\right)^{-\nu}\left[1 - \cos\left(\mathbf{k}\Delta\mathbf{x}_0\right)\right],\\
&= \frac{2C(\nu)l_{2\mathrm{D}}\delta B_{2\mathrm{D}}^2}{\pi t}\left(\frac{\Omega}{vB_0}\right)^2\int_0^\infty \frac{\mathrm{d}k_\perp}{k_\perp^2}\left(1 + k_\perp^2 l_{2\mathrm{D}}^2\right)^{-\nu}\\
&\times\int_0^{2\pi}\mathrm{d}\Psi\;\left[1 - \cos\left(k_\perp\Delta x_0\cos\Psi + k_\perp\Delta y_0\sin\Psi\right)\right].
\end{aligned} \tag{3.60}$$

By using (1.11) for the unperturbed orbit and by using

$$\mathrm{e}^{\pm\mathrm{i}\mathbf{k}\Delta\mathbf{x}} = \sum_{n,m} J_n(W)J_m(W)\,\mathrm{e}^{\pm\mathrm{i}n(\Psi+\Omega t)\mp\mathrm{i}m\Psi} \tag{3.61}$$

and

$$\int_0^{2\pi} d\Psi \, e^{i(n-m)\Psi} = 2\pi \delta_{n,m}, \tag{3.62}$$

we find

$$D_{\mu\mu}(t) = \frac{4C(\nu) l_{2D} \delta B_{2D}^2}{t} \left(\frac{\Omega}{v B_0} \right)^2 \int_0^\infty \frac{dk_\perp}{k_\perp^2} (1 + k_\perp^2 l_{2D}^2)^{-\nu} \times \left[1 - \sum_{n=-\infty}^{\infty} J_n^2(W) \cos(n\Omega t) \right]. \tag{3.63}$$

With

$$\sum_{n=-\infty}^{\infty} J_n^2(W) = 1 \tag{3.64}$$

and

$$1 - \cos(n\Omega t) = 2 \sin^2 (n\Omega t/2), \tag{3.65}$$

one gets

$$D_{\mu\mu}(t) = 8C(\nu) l_{2D} \delta B_{2D}^2 \left(\frac{\Omega}{v B_0} \right)^2 \times \int_0^\infty \frac{dk_\perp}{k_\perp^2} (1 + k_\perp^2 l_{2D}^2)^{-\nu} \sum_{n=-\infty}^{\infty} J_n^2(W) \frac{\sin^2 (n\Omega t/2)}{t}. \tag{3.66}$$

Obviously

$$D_{\mu\mu}(t) \sim \begin{cases} 0 & \text{if } n\Omega t = 2\pi m, \\ t^{-1} & \text{if } n\Omega t \neq 2\pi m, \end{cases} \tag{3.67}$$

where m is an integer number. In Fig. 3.1, we evaluate (3.66) numerically to demonstrate the periodic behavior of $D_{\mu\mu}$. After one gyroperiod, we have $\Delta \delta A_z = 0$ and thus $D_{\mu\mu} = 0$ (see (3.55)). Only between a period where the orbit is not closed, we find nonvanishing pitch-angle diffusion within QLT. In the limit $t \to \infty$, we find $D_{\mu\mu} \to 0$. Thus, the quasilinear parallel mean free path is infinitely large.

3.4.3 Perpendicular Diffusion in Pure 2D Turbulence

Again, we start with the assumption that guiding centers follow magnetic field lines (3.44). In this case, the quasilinear perpendicular displacement of the particle

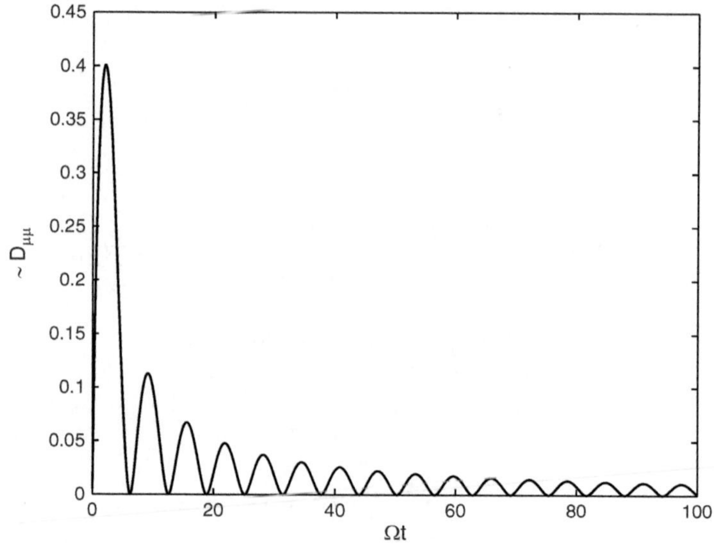

Fig. 3.1 The quasilinear pitch-angle Fokker–Planck coefficient $D_{\mu\mu}$ as a function of Ωt for pure 2D geometry. Obviously, we find $D_{\mu\mu}(t = 2\pi m/(n\Omega)) = 0$ and $D_{\mu\mu}(t \neq 2\pi m/(n\Omega)) \sim t^{-1}$. Thus, a strong subdiffusive behavior of pitch-angle diffusion is obtained within QLT

can be written as

$$\begin{aligned}\Delta x(t) &= \int_0^t \mathrm{d}\tau\, v_x(\tau) = \frac{v\mu}{B_0}\int_0^t \mathrm{d}\tau \delta B_x(\tau),\\ &= \frac{v\mu}{B_0}\int \mathrm{d}^3k \delta B_x(\mathbf{k}) \int_0^t \mathrm{d}\tau\, \mathrm{e}^{\mathrm{i}\mathbf{k}\cdot\mathbf{x}_0(\tau)}.\end{aligned} \tag{3.68}$$

By again assuming homogeneous turbulence, we find for the mean square displacement in the perpendicular direction

$$\left\langle (\Delta x)^2 \right\rangle = \left(\frac{v\mu}{B_0}\right)^2 \int \mathrm{d}^3k P_{xx}(\mathbf{k}) \int_0^t \mathrm{d}\tau_1 \int_0^t \mathrm{d}\tau_2\, \mathrm{e}^{\mathrm{i}\mathbf{k}\cdot[\mathbf{x}(\tau_1)-\mathbf{x}(\tau_2)]} \tag{3.69}$$

and with (2.50)

$$\begin{aligned}\left\langle (\Delta x)^2 \right\rangle &= \left(\frac{v\mu}{B_0}\right)^2 \int_0^\infty \mathrm{d}k_\perp g^{2D}(k_\perp) \int_0^{2\pi} \mathrm{d}\Psi \sin^2\Psi\\ &\times \int_0^t \mathrm{d}\tau_1 \int_0^t \mathrm{d}\tau_2\, \mathrm{e}^{\mathrm{i}k_x[x_0(\tau_1)-x_0(\tau_2)]+\mathrm{i}k_y[y_0(\tau_1)-y_0(\tau_2)]}.\end{aligned} \tag{3.70}$$

To proceed, we use the general relation

$$e^{ik_x(x_0(\tau_1)-x_0(\tau_2))+ik_y(y_0(\tau_1)-y_0(\tau_2))} = \sum_{n,m=-\infty}^{\infty} J_n(W)J_m(W)\, e^{in(\Psi-\Phi_0+\Omega\tau_1)-im(\Psi-\Phi_0+\Omega\tau_2)}, \tag{3.71}$$

with $W = R_L k_\perp \sqrt{1-\mu^2}$. By averaging over the initial gyrophase Φ_0, performing the Ψ-integral, and by using (3.62), we find

$$\begin{aligned}\left\langle(\Delta x)^2\right\rangle &= 2\pi \left(\frac{v\mu}{B_0}\right)^2 \int_0^\infty dk_\perp\, g^{2D}(k_\perp) \\ &\quad \times \sum_{n=-\infty}^{\infty} J_n^2(W) \int_0^t d\tau_1 \int_0^t d\tau_2\, e^{in\Omega(\tau_1-\tau_2)}, \\ &= 2\pi \left(\frac{v\mu}{B_0}\right)^2 \int_0^\infty dk_\perp g^{2D}(k_\perp) \\ &\quad \times \left[J_0^2(W)t^2 + 4\sum_{n=1}^{\infty} J_n^2(W) \frac{\sin^2(n\Omega t/2)}{(n\Omega)^2}\right]. \end{aligned} \tag{3.72}$$

For large enough times ($t \to \infty$), the term proportional to t^2 is dominant and hence

$$\left\langle(\Delta x)^2\right\rangle \approx 2\pi \left(\frac{v\mu t}{B_0}\right)^2 \int_0^\infty dk_\perp\, g^{2D}(k_\perp) J_0^2(W). \tag{3.73}$$

Regardless of the form of the wave spectrum, we find within QLT $\left\langle(\Delta x)^2\right\rangle \sim t^2$. Obviously, quasilinear perpendicular transport in 2D magnetostatic turbulence behaves superdiffusively, or more specific, we find a ballistic motion for all time-scales.

3.5 Quasilinear Transport in the Slab/2D Composite Model

As described in Chap. 2, solar wind turbulence can be well approximated by a superposition of the pure slab and the pure 2D modes. In the slab/2D composite model, all Fokker–Planck coefficients can be written as the sum of the slab coefficient and the 2D coefficient:

$$\begin{aligned} D_{\mu\mu} &= D_{\mu\mu}^{\text{slab}} + D_{\mu\mu}^{2D}, \\ D_\perp &= D_\perp^{\text{slab}} + D_\perp^{2D}. \end{aligned} \tag{3.74}$$

The parallel mean free path is, therefore,

$$\lambda_{\parallel}^{\text{comp}} = \frac{3v}{4} \int_{-1}^{+1} \mathrm{d}\mu \, \frac{(1-\mu^2)^2}{D_{\mu\mu}^{\text{slab}} + D_{\mu\mu}^{\text{2D}}}, \tag{3.75}$$

and the perpendicular mean free path is simply additive

$$\lambda_{\perp}^{\text{comp}} = \lambda_{\perp}^{\text{slab}} + \lambda_{\perp}^{\text{2D}}. \tag{3.76}$$

Therefore, it is straightforward to obtain quasilinear mean free paths by using the results for pure slab and pure 2D:

- For pure slab geometry, quasilinear parallel and perpendicular transport behave diffusively ((3.39) and (3.48)).
- For pure 2D geometry, the parallel mean free path is infinitely large due to the vanishing pitch-angle Fokker–Planck coefficient (see e.g., Fig. 3.1).
- For pure 2D, we find a ballistic motion of the particles in the perpendicular direction (3.73).
- For slab/2D composite geometry, the parallel mean free path is entirely controlled by the slab contribution to the parallel mean free path and we therefore have diffusive behavior.
- Because of the ballistic behavior of the 2D contribution, perpendicular transport also behaves superdiffusively in the composite model.

All results are summarized in Table 3.1. It should be emphasized that these results have been obtained by employing the magnetostatic limit. For dynamical turbulence, however, quasilinear perpendicular diffusion is recovered (see Shalchi and Schlickeiser 2004b). Furthermore, we have $D_{\mu\mu}^{2D} \neq 0$ within these improved turbulence models (Shalchi and Schlickeiser 2004a). In the following, we consider test-particle simulations to check the validity of the quasilinear results.

Table 3.1 Results of QLT obtained for the magnetostatic turbulence model

Turbulence geometry	Transport direction	Behavior of transport
Slab	Parallel	Diffusive
Slab	Perpendicular	Diffusive
2D	Parallel	Superdiffusive
2D	Perpendicular	Superdiffusive
Slab/2D composite	Parallel	Diffusive
Slab/2D composite	Perpendicular	Superdiffusive

Although we find a diffusive behavior of parallel and perpendicular transport in the slab model, we find superdiffusion for pure 2D turbulence geometry. In two-component turbulence, parallel transport behaves diffusively, and perpendicular transport is superdiffusive due to the dominant 2D contribution. In Sect. 3.6, it will be shown that QLT results are not valid for most of the cases considered here

3.6 Test-particle Simulations

In Sects. 3.1–3.5, QLT was applied to calculate parallel and perpendicular diffusion coefficients. Another possibility to obtain transport parameters is to perform test-particle simulations. It should be noted, however, that previous simulations assumed magnetostatic or undamped plasma wave turbulence models (see Table 3.2). In Chap. 8, it will be demonstrated that dynamical turbulence effects are essential if cosmic ray transport is considered for smaller energies. Therefore, simulations are not appropriate for describing transport phenomena for all parameter regimes. Furthermore, for applications such as solar modulation studies, it is useful to have analytical expressions for the mean free paths. Such analytic forms can only be obtained from theoretical investigations. However, test-particle simulations are a very useful tool to test transport theories such as QLT.

One of the first simulations have been performed by Kaiser (1975) and Kaiser et al. (1978), where pitch-angle and parallel spatial diffusion for the magnetostatic slab model have been considered. During the following years, similar and improved test-particle simulation codes have been developed (see Table 3.2). Although some of the simulations are very interesting (e.g., Michalek and Ostrowski 1996 and Michalek 2001, who simulated particle transport in plasma wave turbulence), in this book, only the simulations of Giacalone and Jokipii (1999) and the simulations of Qin (2002), Qin et al. (2002a,b, 2006) are discussed and are compared with QLT and, in Chaps. 4–7, with nonlinear theories.

Table 3.2 Examples of previous test-particle simulations that have been performed to study the propagation of charged particles in magnetic systems

Published in	Turbulence geometry	Temporal behavior
Kaiser (1975) Kaiser et al. (1978)	Slab	Magnetostatic
Owens and Gombosi (1980)	Slab	Magnetostatic
Kóta et al. (1982)	Slab	Magnetostatic
Urch (1984)	Slab	Magnetostatic
Michalek and Ostrowski (1996)	Slab	Undamped plasma waves
Giacalone and Jokipii (1999)	Isotropic/composite	Magnetostatic
Mace et al. (2000)	Slab	Magnetostatic
Michalek (2001)	Isotropic	Undamped plasma waves
Qin et al. (2002a)	Slab	Magnetostatic
Qin et al. (2002b)	Composite	Magnetostatic
Candia and Roulet (2004)	Isotropic	Magnetostatic
Zimbardo et al. (2006)	Anisotropic	Magnetostatic
Qin et al. (2006)	Pure 2D/composite	Magnetostatic

Although the first simulations have explored particle transport in magnetostatic slab turbulence, newer simulations have been done for more realistic models such as the two-component model, isotropic, or anisotropic geometry. Except the simulations of Michalek and Ostrowski (1996) and Michalek (2001), all authors assumed magnetostatic turbulence

3.6.1 The Simulations of Giacalone and Jokipii

In the simulations of Giacalone and Jokipii (1999), two different turbulence geometries have been investigated: the isotropic and the slab/2D composite model. The trajectories of the particles are then obtained by numerically integrating the Lorentz force on each particle (see (1.1)), based on the Burlish–Stoer method (see Press et al. 1986). In all simulations, 2,500 particles are considered, and 50 different, randomly generated realizations of the turbulent fields are used.

In the simulations, the static turbulent magnetic field is derived from the summation over a large number of plane waves with a spherically symmetric propagation direction, which stands for a time-independent Fourier representation in wavenumber space. In the isotropic scenario, all waves have random distributed phases and polarizations, while, in the composite turbulence geometry, 20% slab fluctuations and 80% strictly two-dimensional fluctuations are assumed. The power spectrum of the magnetic fluctuations is limited to wavelengths between $\lambda_{\mathrm{min}} = 10^{-4}$ AU and $\lambda_{\mathrm{max}} = 1$ AU. These parameters λ_{min} and λ_{max} are the same for both the simulations in composite and isotropic turbulence.

3.6.2 The Simulations of Qin

More recently, test-particle simulations have been proposed by Qin (2002) and Qin et al. (2002a, b, 2006). In these simulations, diffusion coefficients have also been calculated by integrating the Newton–Lorentz equation. To obtain the trajectories of 1,000 test-particles, a total number of approximately four million slab modes were selected, and the occurrence of periodic effects was suppressed by taking the box to be 10,000 times larger than the correlation scale of the slab turbulence. The 2D contribution is axisymmetrically distributed in the plane perpendicular to the z axis. The 2D correlation length is taken to be 10% of the slab correlation length $l_{\mathrm{2D}} = 0.1 l_{\mathrm{slab}}$. For the waves in x and y directions, 4096×4096 modes are assumed. Furthermore, in the composite model, where a 20/80 contribution of slab/2D fluctuations is considered, the amplitude of the fluctuating magnetic field is set to $\delta B = B_0$. As in the simulations of Giacalone and Jokipii, a magnetostatic model was applied, and the electric fields were neglected. Qin examined particle transport in pure slab geometry and in the slab/2D composite model.

3.6.3 Confirmation of QLT for Parallel Diffusion in the Slab Model

In the first applications of QLT, a simple magnetostatic slab model without dissipation effects in the spectrum had been assumed (e.g., Jokipii 1966). Therefore,

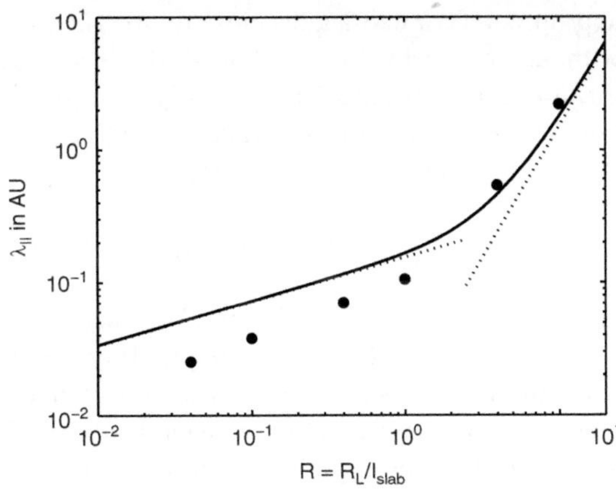

Fig. 3.2 The parallel mean free path, $\lambda_{\parallel}$, for pure slab geometry and moderate turbulence ($\delta B_{\rm slab}^2/B_0^2 = 1$). The simulations (*dots*) are compared with the exact QLT-result (*solid line*) and with the asymptotic approximations of the exact QLT-result (*dotted line*). With the exception of the small discrepancy for low rigidities, QLT agrees well with the simulations

the test-particle codes developed in the 1970s and 1980s of the last century also assumed dissipationless slab turbulence. In Fig. 3.2, the Qin et al. code is applied to determine the parallel mean free path $\lambda_{\parallel}$ as a function of the dimensionless rigidity $R = R_{\rm L}/l_{\rm slab}$. The following parameters have been used to obtain these results:

$$\begin{aligned} \delta B_{\rm slab}/B_0 &= 1, \\ l_{\rm slab} &= 0.03\ {\rm AU}, \\ \nu &= 5/6. \end{aligned} \tag{3.77}$$

The results are compared with the exact QLT result (3.39) and with the asymptotic approximations ((3.41) and (3.42)). Because of the agreement between QLT and simulations, several authors applied QLT also for describing particle propagation for nonslab models and for perpendicular transport. In the following, it is demonstrated that there are at least three problems associated with the application of QLT.

3.7 The Three Problems of QLT

Besides the so-called 90°-problem, QLT is also incorrect for perpendicular transport and for parallel diffusion in nonslab models (geometry problem). In the following, these three problems are discussed.

3.7.1 The 90°-Scattering Problem

A short time after QLT had been derived, it was noted by several authors (e.g., Klimas and Sandri 1971, 1973, Völk 1973, Kaiser et al. 1973) that QLT is not appropriate for pitch-angles close to 90° (i.e., $\mu = 0$). As long as a wave spectrum without steep behavior is assumed, the inaccuracy of QLT is small if the parallel mean free path is calculated (see Fig. 3.2). For steep spectra, however, we find an infinitely large quasilinear parallel mean free path that we can easily understand. As shown in (3.37), we have, within QLT

$$D_{\mu\mu}(\mu \to 0) \sim \mu^{2\nu-1}, \tag{3.78}$$

and thus the integral for the parallel mean free path (3.2) is not convergent for $2\nu \geq 2$: $\lambda_{\parallel}(2\nu \geq 2) \to \infty$. Jokipii (1968) proposed a different derivation of $\lambda_{\parallel}$, still within the quasilinear approximation, which makes certain a priori assumptions about the pitch-angle distribution and leads to a convergent expression for $\lambda_{\parallel}$, irrespective of the type of power spectrum. Earl (1973) correctly argued that, in general, this procedure is not justified. In any case, it appears only to cover up the true problem. The basic inconsistency lies with the pitch angle diffusion coefficient and with the assumption that a particle follows its unperturbed orbit, irrespective of its parallel velocity. In reality, as time proceeds, the particle orbit will deviate more and more from the unperturbed helical trajectory.

In the heliosphere, steep wave spectra can be seen at short turbulence scales (dissipation range). As shown in Fig. 2.3, the dissipation range spectral index is approximately equal to 3. To explore the failure of QLT, a test-particle code can be used to calculate the pitch-angle Fokker–Planck coefficient. In Fig. 3.3, we compare QLT with the simulations performed by Qin (2002) for the following set of parameters:

$$\begin{aligned} \delta B_{\mathrm{slab}}/B_0 &= 0.05, \\ R &= 0.0363\,\mathrm{AU}, \\ \nu &= 5/6. \end{aligned} \tag{3.79}$$

As illustrated by Fig. 3.3, we already find disagreement between QLT and simulations for very weak turbulence. For strong turbulence as in the solar wind ($\delta B \approx B_0$), nonlinear effects at 90° are expected to be much stronger (see Chap. 6).

After the failure of QLT had been discovered, two approaches have been proposed to prevent the parallel mean free path from becoming infinitely large for realistic spectra:

1. The quasilinear approximation can be replaced by a more accurate transport theory. Mainly during the 1970s, several nonlinear theories had been developed to solve the 90°-problem, some of which will be discussed in Chaps. 4 and 6.
2. The magnetostatic model can be replaced by a nonstatic model:

   a. For example, Schlickeiser (2002) has shown that, if the effect of propagating plasma waves is combined with QLT, nonvanishing pitch-angle scattering at 90° can be obtained.

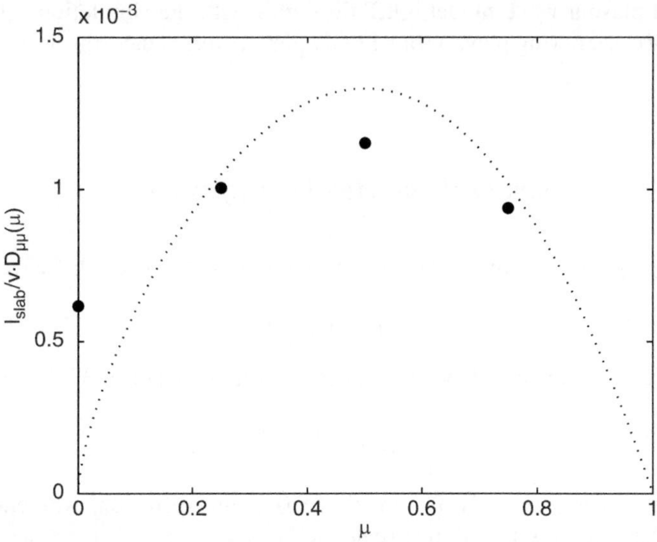

Fig. 3.3 The pitch-angle Fokker–Planck coefficient $D_{\mu\mu}$. Shown are the results of QLT (*dotted line*) and simulations (*dots*). The 90°-problem is represented by the disagreement of these two results for small values of μ

b. Other authors included the effect of dynamical turbulence (e.g., Bieber et al. 1994). In the case of the damping model of dynamical turbulence (DT model, see Chap. 2), for instance, the quasilinear resonance function becomes a Breit–Wigner function

$$\begin{aligned} R_n^{\mathrm{QLT,MS}} &= \pi\delta(v\mu k_\parallel + n\Omega) \\ \to R_n^{\mathrm{QLT,DT}} &= \frac{\alpha v_{\mathrm{A}} k}{(\alpha v_{\mathrm{A}} k)^2 + (v\mu k_\parallel + n\Omega)^2}. \end{aligned} \tag{3.80}$$

For pitch-angle scattering and slab turbulence, only $n = \pm 1$ is relevant. Thus, we find at 90° (i.e., $\mu = 0$)

$$R_n^{\mathrm{QLT,DT}}(\mu = 0) = \pi\delta(\pm\Omega) = 0 \to \frac{\alpha v_{\mathrm{A}} k}{(\alpha v_{\mathrm{A}} k)^2 + \Omega^2} \neq 0 \tag{3.81}$$

as long as $\alpha \neq 0$.

Within both approaches, we no longer find an infinitely large parallel mean free path. However, the second approach to the problem does not really solve the problem itself, namely the invalidity of QLT at 90°. It merely prevents the parallel mean free path from becoming infinitely large. Therefore, one expects that, even for more realistic turbulence models, quasilinear parallel mean free paths are not valid. For example, Tautz et al. (2006b) have demonstrated that, within the undamped plasma wave model, QLT disagrees with the simulations of Michalek (2001) if 90°-scattering plays a role. For dynamical turbulence, the accuracy of QLT is unclear.

3.7.2 The Problem of Perpendicular Diffusion

By employing QLT, we find for magnetostatic slab turbulence a diffusive result (see (3.48)):

$$\left\langle (\Delta x)^2 \right\rangle \sim t. \tag{3.82}$$

For pure 2D turbulence and the slab/2D composite model (see (3.73)), we find

$$\left\langle (\Delta x)^2 \right\rangle \sim t^2. \tag{3.83}$$

The first result corresponds to normal (Markovian) diffusion, whereas the latter shows ballistic motion for all time scales.

To explore the validity of QLT, several test-particle simulations have been performed in the past (e.g., Giacalone and Jokipii 1999, Qin et al. 2002a, b). Qin et al. (2002a), for instance, assumed slab geometry and they found subdiffusion (see Fig. 3.4). More specific, it was found that

$$\left\langle (\Delta x)^2 \right\rangle \sim \sqrt{t}. \tag{3.84}$$

Already in the 1970s of the last century, some theoretical evidence for subdiffusion had been discovered (see e.g., Urch 1977). Some years later, Jokipii et al. (1993) formulated the theorem on reduced dimensionality, where it has been claimed that, for pure slab geometry, the perpendicular mean free path must be zero as particles are tied to magnetic lines of force. Kóta and Jokipii (2000) proposed a compound diffusion model, which states that parallel diffusion has a strong influence if the perpendicular mean free path is calculated. Under the assumption of diffusive field line wandering, these authors derived that

$$\left\langle (\Delta x)^2 \right\rangle = 2\kappa_{\mathrm{FL}} \sqrt{2t\kappa_{\parallel}} \sim \sqrt{t}, \tag{3.85}$$

which is smaller than the standard diffusion result, for which we have $\left\langle (\Delta x)^2 \right\rangle = 2t\kappa_{xx}$, and, therefore, is fundamentally non-Markovian. Obviously, these results are in agreement with test-particle simulations (in both we find $\left\langle (\Delta x)^2 \right\rangle \sim \sqrt{t}$). On the other hand, it has been claimed in several observational papers that $\lambda_{\perp}/\lambda_{\parallel} \approx 0.01$ and, in more recent articles (Dwyer et al. 1997), it has been stated that $\lambda_{\perp}/\lambda_{\parallel}$ can approach or exceed unity. To achieve a more complete understanding of the physics of perpendicular transport, more test-particle simulations have been performed, mainly for nonslab models. An interesting and important result has been obtained: perpendicular diffusion is recovered for nonslab models.

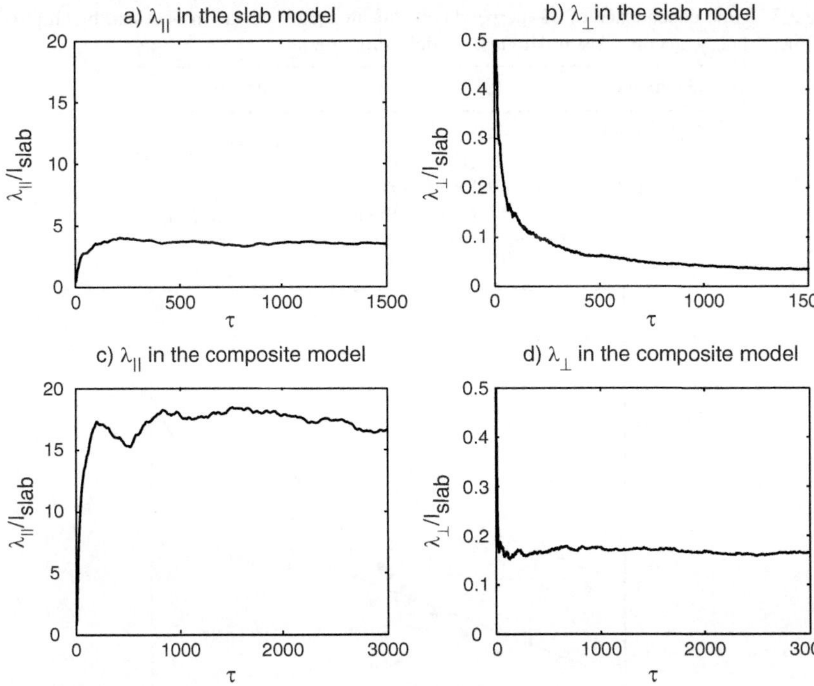

Fig. 3.4 Time dependent diffusion coefficients from test-particle simulations. For all four figures, the following parameters have been used: $R = 1$, $\nu = 5/6$, $\delta B = B_0$; (**a**) Shown is the parallel mean free path as a function of the dimensionless time defined as $\tau = vt/l_{\text{slab}}$ for slab geometry. Clearly, the particle reaches diffusive behavior after a certain time-scale. (**b**) The perpendicular mean free path as a function of time. In the slab model, we find subdiffusion. (**c**) The parallel mean free path for 20%/80% composite geometry. Again the parallel transport behaves diffusively. (**d**) Perpendicular transport in the composite model: diffusion is recovered

Furthermore, it has been demonstrated in simulations (see e.g., Fig. 3.4) that

$$\frac{\lambda_\perp}{\lambda_\parallel} \approx 0.01, \tag{3.86}$$

in agreement with observations. As summarized in Table 3.3, all these results are in disagreement with QLT. Obviously, QLT fails for perpendicular transport.

3.7.3 The Geometry Problem

Clearly, QLT is incorrect for perpendicular transport, independent of the turbulence model. Furthermore, QLT cannot be applied if the parallel mean free path is calculated for steep wave spectra. In the previous years, these two problems have been

Table 3.3 Time-dependence of the perpendicular mean square deviation obtained by employing quasilinear theory, in comparison with test-particle simulations

Geometry	QLT	Simulations
Slab model	$\langle(\Delta x)^2\rangle \sim t$ (Diffusion)	$\langle(\Delta x)^2\rangle \sim \sqrt{t}$ (Subdiffusion)
Composite model	$\langle(\Delta x)^2\rangle \sim t^2$ (Superdiffusion)	$\langle(\Delta x)^2\rangle \sim t$ (Diffusion)

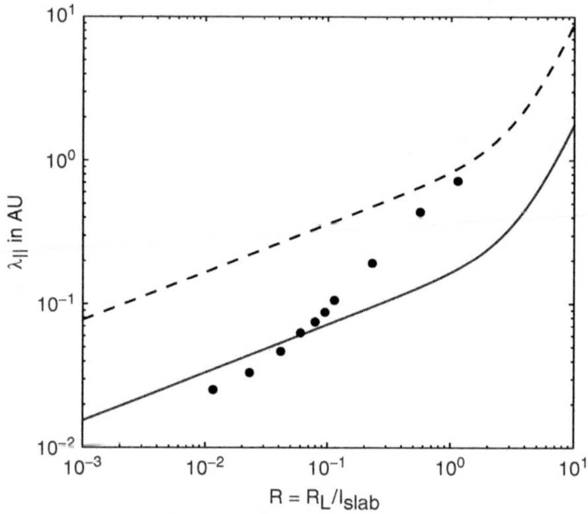

Fig. 3.5 Simulations for 20% slab/80% 2D composite geometry (*dots*) in comparison with QLT-results for the parallel mean free path. Shown are QLT results for pure slab (*dotted line*) and 20% slab/80% 2D composite geometry (*dashed line*) as a function of the dimensionless rigidity $R = R_L/l_{slab}$ (R_L is the gyroradius and l_{slab} is the slab bendover scale)

well known and, therefore, QLT has only been applied to describe parallel transport for flat spectra. For certain turbulence models, however, singularities have been discovered to be associated with the 90°-problem. For instance, for isotropic turbulence it can easily be derived (see e.g., Tautz et al. 2006a) that $D_{\mu\mu}(\mu \to 0) \sim \mu^{-s}$ (s is the turbulence spectral index, see Chap. 2) and the parallel mean free path calculated with (3.2) becomes infinitely large for $s \geq 1$. Already for the realistic value $s \approx 5/3$, the 90°-problem generates an infinitely large parallel spatial diffusion coefficient.

For the slab/2D composite model, however, 90°-scattering is not important as long as flat spectra ($s \leq 2$) are assumed. In Fig. 3.5, simulations are compared with QLT results for the parameter set listed in Table 3.4. Clearly, a disagreement can be seen. Obviously, QLT is incorrect for parallel diffusion even in cases where the 90°-problem does not occur. The first time that QLT has been compared for composite geometry with simulations was in Giacalone and Jokipii (1999). The authors stated that the disagreement between QLT and their simulations is small. However, their

Table 3.4 Parameters used for numerical and theoretical calculations

Parameter	Symbol	Value
Inertial range spectral index	2ν	$5/3$
Slab bendover scale	$l_{\rm slab}$	0.030 AU
2D bendover scale	$l_{\rm 2D}$	$0.1\ l_{\rm slab}$
Turbulence strength	$\delta B/B_0$	1
Slab fraction	$\delta B^2_{\rm slab}$	$0.2\ \delta B^2$
2D fraction	$\delta B^2_{\rm 2D}$	$0.8\ \delta B^2$

The values should be appropriate for solar wind parameters at 1 AU heliocentric distance

statement is based on a wrong comparison (unfortunately, Giacalone and Jokipii compared their simulations for isotropic and composite geometry with QLT-slab results). Some years later, Minnie (2002) noticed this error and presented a correct comparison similar to Fig. 3.5. Another 2 years later, Shalchi et al. (2004b) revisited this third problem of cosmic ray transport theory and called it the geometry problem.

Because the slab/2D composite model seems to be an appropriate model to represent interplanetary turbulence (see Bieber et al. 1994, 1996), a correct theory for describing parallel diffusion in composite turbulence is essential. Such a nonlinear theory will be described in Chap. 5, namely the weakly nonlinear theory.

An interesting feature of the simulations shown in Fig. 3.5 is the steep rigidity dependence. Although, within QLT, we find $\lambda_\parallel \sim R^{1/3}$ for a Kolmogorov spectrum ($\nu = 5/6$), we find a much steeper rigidity dependence in the Qin simulations and in other simulations (Giacalone and Jokipii 1999, Michalek 2001). As discussed in Chaps. 1 and 8, this behavior is more in agreement with observations of galactic cosmic rays. It should be emphasized that, for isotropic or anisotropic turbulence, both the 90° and the geometry problem occur. Therefore, one comes to the conclusion that QLT is only applicable for slab geometry and flat wave spectra.

Chapter 4
The Nonlinear Guiding Center Theory

In the previous chapter, it has been demonstrated that, in most cases, a quasilinear description of cosmic ray transport is not accurate and is invalid for at least some parameter regimes or turbulence models. In this chapter, previous nonlinear transport theories are discussed. As a representative of the theories that have been developed in the 1970s of the previous century, we discuss the nonlinear closure approximation that has been derived to describe 90°-scattering and to solve the problem of perpendicular transport. Also the Bieber and Matthaeus model is discussed that has been developed to describe perpendicular diffusion and drifts of charged particles. More recently, the so-called nonlinear guiding center theory has been formulated to describe perpendicular diffusion, in agreement with test-particle simulations. After discussing the basic assumptions of this advanced diffusion theory, we consider a comparison with computer simulations and we discuss analytical forms of the perpendicular mean free path.

4.1 The Nonlinear Closure Approximation

Owens (1974) considered Liouville's equation for the conservation of the cosmic ray particle density in phase-space, with the magnetic field as the dominant force. Although QLT ignores all nonlinear terms in this equation, the method of Owens includes them in the nonlinear closure approximation (NCA). This approximation is intended to give only a semi-quantitative indication of the effect of the nonlinear terms in the theory. According to Owens, the nonlinear closure approximation allows one to draw more conclusions than in earlier methods (e.g., Dupree 1966, 1967) by explicitly retaining the effects of the nonlinear terms in all equations, although in an approximate and somewhat heuristic way.

The solution deduced by Owens (see also Owens 1973) arises from the imposition of initial conditions decay exponentially as

$$\exp\left(-t/\tau_s\right),$$

A. Shalchi, *Nonlinear Cosmic Ray Diffusion Theories*,
Astrophysics and Space Science Library,
DOI: 10.1007/978-3-642-00309-7_4,

where t is the time measured from the initial time and τ_s is the scattering time. This model was the first to show that the initial-value terms damp out with time; all previous derivations, including standard QLT and the nonlinear theories based on Dupree's work (Dupree 1966, 1967), merely ignored the initial conditions. As the nonlinear closure approximation shows, the physically expected decay of the initial conditions is due to higher-order terms in Liouville's equation, and the decay is exponential in time, with the decay constant given by the scattering time.

Instead of the sharp resonance function of QLT, the nonlinear closure model gives the form

$$\begin{aligned} R_n^{\mathrm{QLT}} &= \pi\delta\left(k_\parallel v_\parallel + n\Omega\right) \\ \to R_n^{\mathrm{NCA}} &= \Re\mathfrak{e} \int_0^\infty \mathrm{d}t\, \mathrm{e}^{\mathrm{i}(k_\parallel v\mu + n\Omega)t}\, \mathrm{e}^{-t/\tau_\mathrm{s}} = \frac{\tau_\mathrm{s}^{-1}}{\left(k_\parallel v_\parallel + n\Omega\right)^2 + \tau_\mathrm{s}^{-2}}. \end{aligned} \tag{4.1}$$

QLT gives a delta function or infinitely narrow resonance, while NCA gives a Breit–Wigner resonance function with a width of τ_s^{-1}. Physically, the broadening of the cyclotron resonance obtained from NCA is easily understood. The resonances occur due to the coupling between the cosmic ray particle and the magnetic fluctuations. The resonant wavenumber is given by $k_\parallel v_\parallel = n\Omega$. In QLT, the particles are assumed to remain on the unperturbed trajectories, and hence the interaction takes place for an infinitely long time and the resonance width becomes infinitely narrow. In practice, of course, a given particle can interact with a given wavelength of the fluctuations only as long as it is not scattered to another region of phase space. The time that a particle can spend in phase with the fluctuations is, therefore, of the order of τ_s.

Owens argued that the parallel diffusion coefficient is always larger than the perpendicular diffusion coefficient. Thus, parallel diffusion is the dominant process and controls the scattering time

$$\tau_\mathrm{s} = \lambda_\parallel / v. \tag{4.2}$$

Equation (4.2), in combination with the NCA resonance function of (4.1), is the basis of Owens theory.

4.1.1 The Results of the NCA

Here, the consequences of the NCA onto parallel and perpendicular diffusion are considered and discussed for pure slab geometry. With the resonance function from (4.1), we find the following form of the pitch-angle Fokker–Planck coefficient (see (3.34)):

$$D_{\mu\mu} = \frac{\pi\Omega^2(1-\mu^2)}{B_0^2} \int_{-\infty}^{\infty} \mathrm{d}k_\parallel\, g^{\mathrm{slab}}(k_\parallel) \sum_{n=\pm 1} \frac{\tau_\mathrm{s}^{-1}}{\tau_\mathrm{s}^{-2} + (k_\parallel v\mu + n\Omega)^2} \tag{4.3}$$

and especially at 90° ($\mu = 0$):

$$D_{\mu\mu}(\mu = 0) = \frac{2\pi \Omega^2}{B_0^2} \int_{-\infty}^{\infty} dk_\parallel \, g^{\text{slab}}(k_\parallel) \frac{\tau_s^{-1}}{\tau_s^{-2} + \Omega^2}. \tag{4.4}$$

By using

$$\int_{-\infty}^{\infty} dk_\parallel \, g^{\text{slab}}(k_\parallel) = \frac{\delta B_{\text{slab}}^2}{4\pi} \tag{4.5}$$

and (4.2), we find

$$D_{\mu\mu}(\mu = 0) = \frac{\Omega^2 \delta B_{\text{slab}}^2}{2B_0^2} \frac{v/\lambda_\parallel}{(v/\lambda_\parallel)^2 + \Omega^2}. \tag{4.6}$$

This result is further discussed in the next paragraph, where we compare this result with simulations.

Similar to the previous steps, we can also determine the Fokker–Planck coefficient of perpendicular diffusion. By combining (3.29) with the magnetostatic slab model and (4.1), we find

$$D_{xx} = 4\pi \frac{v^2 \mu^2}{B_0^2} \int_0^{\infty} dk_\parallel \, g^{\text{slab}}(k_\parallel) \frac{\tau_s^{-1}}{(k_\parallel v \mu)^2 + \tau_s^{-2}}. \tag{4.7}$$

With the spectrum from (2.33), one gets

$$D_{xx} = 2C(\nu) l_{\text{slab}} v^2 \mu^2 \frac{\delta B_{\text{slab}}^2}{B_0^2} \int_0^{\infty} dk_\parallel \, (1 + k_\parallel^2 l_{\text{slab}}^2)^{-\nu} \frac{\tau_s^{-1}}{(k_\parallel v \mu)^2 + \tau_s^{-2}}, \tag{4.8}$$

and for the perpendicular spatial diffusion coefficient (see (1.95)),

$$\kappa_\perp = \tau_s^{-1} C(\nu) l_{\text{slab}} v^2 \frac{\delta B_{\text{slab}}^2}{B_0^2} \int_0^{\infty} dk_\parallel \, (1 + k_\parallel^2 l_{\text{slab}}^2)^{-\nu} \times \int_{-1}^{+1} d\mu \, \frac{\mu^2}{(k_\parallel v \mu)^2 + \tau_s^{-2}}. \tag{4.9}$$

With (4.2), by using the integral transformation $x = k_\parallel l_{\text{slab}}$ and by using $\lambda_\perp = 3\kappa_\perp / v$, we deduce for the ratio of the perpendicular and parallel mean free path

$$\frac{\lambda_\perp}{\lambda_\parallel} = 6C(\nu) \frac{\delta B_{\text{slab}}^2}{B_0^2} \int_0^{\infty} dx \, (1 + x^2)^{-\nu} \int_{-1}^{+1} d\mu \, \frac{\mu^2}{1 + \mu^2 x^2 \lambda_\parallel^2 / l_{\text{slab}}^2}. \tag{4.10}$$

The integral over μ can be solved and we obtain

$$\frac{\lambda_\perp}{\lambda_\parallel} = 6C(\nu)\frac{\delta B_{\text{slab}}^2}{B_0^2}\int_0^\infty dx\ (1+x^2)^{-\nu} \times \left[\left(\frac{\lambda_\parallel x}{l_{\text{slab}}}\right)^{-2} - \left(\frac{\lambda_\parallel x}{l_{\text{slab}}}\right)^{-3}\arctan\left(\frac{\lambda_\parallel x}{l_{\text{slab}}}\right)\right]. \tag{4.11}$$

A numerical investigation of this result would be straightforward. In this book, we consider only analytical results for two extreme cases.

4.1.1.1 The Limit $\lambda_\parallel / l_{\text{slab}} \to 0$

In this limit, we find

$$\left[\left(\frac{\lambda_\parallel x}{l_{\text{slab}}}\right)^{-2} - \left(\frac{\lambda_\parallel x}{l_{\text{slab}}}\right)^{-3}\arctan\left(\frac{\lambda_\parallel x}{l_{\text{slab}}}\right)\right] \to \frac{1}{3} \tag{4.12}$$

and, by using

$$\int_0^\infty dx\ (1+x^2)^{-\nu} = \frac{1}{4C(\nu)}, \tag{4.13}$$

we finally obtain

$$\frac{\lambda_\perp}{\lambda_\parallel} = \frac{1}{2}\left(\frac{\delta B_{\text{slab}}}{B_0}\right)^2. \tag{4.14}$$

This nonvanishing result can be interpreted as a diffusive behavior of perpendicular transport.

4.1.1.2 The Limit $\lambda_\parallel / l_{\text{slab}} \to \infty$

In this case, we can split the integral of (4.11)

$$\frac{\lambda_\perp}{\lambda_\parallel} \approx 6C(\nu)\frac{\delta B_{\text{slab}}^2}{B_0^2}\left\{\int_0^1 dx\ [\ldots] + \int_1^\infty dx\ x^{-2\nu}\,[\ldots]\right\}. \tag{4.15}$$

It can easily be demonstrated that the second contribution can be neglected. The first term can be expressed by a hypergeometric function

$$\frac{\lambda_\perp}{\lambda_\parallel} = 6C(\nu)\frac{\delta B_{\text{slab}}^2}{B_0^2}\frac{l_{\text{slab}}^2}{\lambda_\parallel^2}\left[{}_2F_1\left(-\frac{1}{2}, 1, \frac{3}{2}, -\frac{\lambda_\parallel^2}{l_{\text{slab}}^2}\right) - 1\right], \tag{4.16}$$

which, in the limit $\lambda_\parallel / l_{\text{slab}} \to \infty$, can be approximated by the asymptotic expansion, and we find

$$\lambda_\perp = \frac{3\pi}{2} C(\nu) l_{\text{slab}} \left(\frac{\delta B_{\text{slab}}}{B_0} \right)^2 = \lambda_{\perp,\text{FLRW}}, \tag{4.17}$$

which is the well known quasilinear result (see (3.48)). In the next paragraph, the results from (4.14) and (4.17) are further discussed and compared with simulations.

4.1.2 *Test of the NCA by Comparing it with Simulations*

It is interesting to test the NCA by comparing the results of the previous paragraphs with simulations. First, we consider the dimensionless pitch-angle Fokker–Planck coefficient at 90°, which can easily be derived from (4.6)

$$\tilde{D}_{\mu\mu}(\mu = 0) = \frac{l_{\text{slab}}}{v} D_{\mu\mu}(\mu = 0) = \frac{1}{2} \left(\frac{\delta B_{\text{slab}}}{B_0} \right)^2 \frac{l_{\text{slab}}/\lambda_\parallel}{(R l_{\text{slab}}/\lambda_\parallel)^2 + 1}. \tag{4.18}$$

In the test-particle run presented in Sect. 3.5, we used $R = 0.0363$, $\delta B/B_0 = 0.05$, and $l_{\text{slab}} = 0.03$ AU. From the simulations, we can also determine the parallel mean free path which, for this parameter set, is $\lambda_\parallel \approx 20.0$ AU. By applying (4.18), we find within the NCA $\tilde{D}_{\mu\mu}(\mu = 0) \approx 1.83 \times 10^{-6}$. According to Fig. 3.3, we find in the simulations $\tilde{D}_{\mu\mu}(\mu = 0) \approx 0.6 \times 10^{-3}$, which disagrees with the NCA.

Now we consider the problem of perpendicular diffusion. We note that we obtain finite values for the perpendicular diffusion coefficient within the NCA (see (4.14) and (4.17)). These results have to be interpreted as a diffusive behavior of perpendicular transport. In simulations, however, we find subdiffusion for slab geometry. Therefore, we come to the conclusion that the NCA can solve neither the 90°-problem nor the problem of perpendicular transport of cosmic rays.

4.2 The Bieber and Matthaeus Model

An approach to solve the problem of perpendicular diffusion and to calculate the off-diagonal elements of the diffusion tensors (κ_{xy}, κ_{yx}) was presented by Bieber and Matthaeus (BAM; 1997). In the BAM model, the TGK formula

$$\kappa_{ij} = \int_0^\infty dt \, \langle v_i(t) v_j(0) \rangle \tag{4.19}$$

is used to determine diffusion coefficients. Obviously, the diffusion coefficients are known if the velocity correlation functions

$$V_{ij}(t) = \langle v_i(t) v_j(0) \rangle \tag{4.20}$$

are known. The basic idea of the BAM theory is to replace the functions $V_{ij}(t)$ by appropriate models. In the following, the BAM theory is explained and applied for perpendicular transport and drifts.

4.2.1 The Basic Formulas of the BAM Theory

According to Bieber and Matthaeus (1997), the following statements allow the development of a transport theory for perpendicular diffusion: if there are no turbulent fields ($\delta B_i = 0$) and if the background field is uniform in the z direction ($\mathbf{B}_0 = B_0 \mathbf{e}_z$), the particle trajectory is a perfect helix along which the particle moves at constant speed, while the perpendicular velocity components oscillate at the angular frequency Ω, so that $V_{xx}(t) = V_{yy}(t) \sim \cos(\Omega t)$, $V_{xy}(t) = -V_{yx}(t) \sim \sin(\Omega t)$, and $V_{zz}(t) \sim 1$.

If we consider the perturbed motion ($\delta B_i \neq 0$), the fluctuations cause the particle to wander in pitch-angle and gyrophase. In this case, the particle *forgets* its formerly perfect helical trajectory. After one gyroperiod, the velocity correlation is assumed to fall eventually to zero at large times. The *rate of forgetting* the helical trajectory can be modeled by a simple decorrelation rate $\omega_\perp$, and a reasonable form for the transverse correlation functions is $V_{xx} = V_{yy} \sim \cos(\Omega t)\,\mathrm{e}^{-\omega_\perp t}$. The cross-correlation experiences the same kind of decorrelation, so we can adopt $V_{yx} = -V_{xy} \sim \sin(\Omega t)\,\mathrm{e}^{-\omega_{\mathrm{AS}} t}$.

Similarly, there is a decorrelation rate $\omega_\parallel$ associated with the parallel velocity, but it does not contain an oscillatory term: $V_{zz} \sim \mathrm{e}^{-\omega_\parallel t}$. In addition, one could assume that the scattering occurs with sufficient randomness, then the particle distribution becomes nearly isotropic within the timescales of interest. By assembling these assumptions, we find the following models for the velocity correlation functions

$$\begin{aligned} V_{xx}(t) = V_{yy}(t) &= \frac{v^2}{3} \cos(\Omega t)\,\mathrm{e}^{-\omega_\perp t}, \\ V_{yx}(t) = -V_{xy}(t) &= \frac{v^2}{3} \sin(\Omega t)\,\mathrm{e}^{-\omega_\perp t}, \\ V_{zz}(t) &= \frac{v^2}{3}\,\mathrm{e}^{-\omega_\parallel t}. \end{aligned} \tag{4.21}$$

Here, Bieber and Matthaeus assumed that $\omega_{\mathrm{AS}} = \nu_\perp$. These models can easily be combined with (4.19) to find, within the BAM theory,

$$\begin{aligned} \kappa_\perp = \kappa_{xx} = \kappa_{yy} &= \frac{v^2}{3} \frac{\omega_\perp}{\omega_\perp^2 + \Omega^2}, \\ \kappa_{\mathrm{AS}} = \kappa_{yx} = -\kappa_{xy} &= \frac{v^2}{3} \frac{\Omega}{\omega_\perp^2 + \Omega^2}, \\ \kappa_\parallel = \kappa_{zz} &= \frac{v^2}{3} \frac{1}{\omega_\parallel}. \end{aligned} \tag{4.22}$$

The latter equation can be written as

$$\omega_{\parallel} = \frac{v^2}{3\kappa_{\parallel}} = \frac{v}{\lambda_{\parallel}}, \tag{4.23}$$

where the parallel mean free path $\lambda_{\parallel}$ has been used. By resubstituting (4.23) into (4.21), we find

$$V_{zz}(t) = \frac{v^2}{3}\,\mathrm{e}^{-vt/\lambda_{\parallel}}. \tag{4.24}$$

By introducing the time-scale $\tau = \omega_{\perp}^{-1}$, we find for the perpendicular spatial diffusion coefficient

$$\kappa_{\perp} = \frac{vR_{\mathrm{L}}}{3}\frac{\Omega\tau}{1+(\Omega\tau)^2} \tag{4.25}$$

and for the antisymmetric term

$$\kappa_{\mathrm{AS}} = \frac{vR_{\mathrm{L}}}{3}\frac{(\Omega\tau)^2}{1+(\Omega\tau)^2}. \tag{4.26}$$

Bieber and Matthaeus noted correctly that this set of equations is formally the same as for hard-sphere scattering in a magnetized medium (Gleeson 1969), in which case the parameter τ is the scattering time. A similar structure has been found for collisional diffusion coefficients in a thermal equilibrium plasma by Balescu et al. (1994).

To further evaluate the perpendicular diffusion coefficient, Bieber and Matthaeus (1997) used the assumption that field line random walk is the source of decorrelation of orbits. The average perpendicular deflection of the particles after movement along the mean field by a distance z is $\delta x(z) = \sqrt{2\kappa_{\mathrm{FL}}z}$, where κ_{FL} is the diffusion coefficient associated with the field line random walk (Jokipii and Parker 1969; Matthaeus et al. 1995; see also Chap. 2). To proceed, Bieber and Matthaeus assumed that the perpendicular velocity components decorrelate over the scale z_{c}, corresponding to an average field line deflection by a gyroradius R_{L}:

$$z_{\mathrm{c}} = \frac{R_{\mathrm{L}}^2}{\kappa_{\mathrm{FL}}} \tag{4.27}$$

and, therefore,

$$\tau \sim \frac{z_{\mathrm{c}}}{v} = \frac{R_{\mathrm{L}}^2}{v\kappa_{\mathrm{FL}}}. \tag{4.28}$$

By using $R_{\mathrm{L}} = v/\Omega$, we find

$$\Omega\tau = \frac{2}{3}\frac{R_{\mathrm{L}}}{\kappa_{\mathrm{FL}}}, \tag{4.29}$$

which is, in effect, the decorrelation time in units of the gyroperiod. The factor $2/3$ in (4.29) is chosen to agree with QLT in the appropriate limit (see below). If (4.29) is combined with (4.22) and (4.21), we can determine the functions V_{xx} and V_{yy} as well as the corresponding spatial diffusion coefficients.

4.2.2 *Results of the BAM Theory for Slab Geometry*

According to (4.25), we have for the perpendicular mean free path

$$\lambda_{\perp} = \frac{3}{v}\kappa_{\perp} = R_{\mathrm{L}} \frac{\Omega\tau}{1 + (\Omega\tau)^2}. \tag{4.30}$$

To proceed, we assume magnetostatic slab turbulence and the standard spectrum of (2.33). In this case, field line wandering behaves diffusively and the field line diffusion coefficient is (see (2.89))

$$\kappa_{\mathrm{FL}} = l_{\mathrm{c,slab}} \left(\frac{\delta B_x}{B_0} \right)^2 \equiv \pi C(\nu) l_{\mathrm{slab}} \left(\frac{\delta B_{\mathrm{slab}}}{B_0} \right)^2 \tag{4.31}$$

and thus (see (4.29))

$$\Omega\tau = \frac{2R}{3\pi C(\nu)} \left(\frac{B_0}{\delta B_{\mathrm{slab}}} \right)^2. \tag{4.32}$$

It is straightforward to approximate (4.30) for small and large values of $\Omega\tau$:

$$\begin{aligned} \lambda_{\perp}(\Omega\tau \ll 1) &= R_{\mathrm{L}}\Omega\tau = \frac{2R^2}{3\pi C(\nu)} l_{\mathrm{slab}} \left(\frac{B_0}{\delta B_{\mathrm{slab}}} \right)^2 \\ \lambda_{\perp}(\Omega\tau \gg 1) &= \frac{R_{\mathrm{L}}}{\Omega\tau} = \frac{3\pi}{2} C(\nu) l_{\mathrm{slab}} \left(\frac{\delta B_{\mathrm{slab}}}{B_0} \right)^2 \end{aligned} \tag{4.33}$$

The second formula corresponds to the result obtained by employing QLT. Obviously the quasilinear result can be deduced from the BAM theory by assuming smallness of the parameter $\Omega\tau$. According to (4.29), $\Omega\tau$ is small for large cosmic ray energies (or gyroradii) and for weak turbulence (and therefore small field line diffusion coefficients).

In both cases, we have $\lambda_{\perp} \neq 0$, which has to be interpreted as diffusive behavior. As demonstrated in Chap. 3, however, we find subdiffusion for the magnetostatic slab model. Obviously the BAM model is incorrect at least for slab geometry. The reason for the failure of the BAM theory can easily be understood: in (4.21), exponential forms of the perpendicular velocity correlation functions have been assumed. These forms are in disagreement with the nonexponential velocity correlation function that will be systematically derived in Chap. 7 by using the extended nonlinear guiding center approach developed by Shalchi (2006b). Some further aspects of velocity correlation functions and their theoretical derivation are discussed in Shalchi and Döring (2007).

4.2.3 The BAM Theory for Slab/2D Composite Geometry

For slab/2D composite geometry, the analytical expression of (4.31) for the field line diffusion coefficient no longer holds. Matthaeus et al. (1995) developed a nonperturbative theory (see (2.113)) for magnetic field line diffusion, stating that

$$\kappa_{FL} = \frac{1}{2}\kappa_{slab} + \frac{1}{2}\sqrt{\kappa_{slab}^2 + 4\kappa_{2D}^2}, \tag{4.34}$$

with the slab field line diffusion coefficient of (4.31) and with the 2D field line coefficient of (2.111). It seems to be straightforward to combine the results for the field line diffusion coefficient with (4.29) and (4.25) to determine the perpendicular mean free path for the two component model. However, as demonstrated in Chap. 2, we find for the standard spectrum of (2.51) an infinitely large parameter κ_{2D}. To prevent this singularity, Matthaeus et al. (1995) introduced another length-scale, namely the so-called ultra-scale $l_{2D,ultra}$. This technique leads to the form

$$\kappa_{2D} = l_{2D,ultra}\frac{\delta B_{2D}}{B_0}. \tag{4.35}$$

By combining (4.35) with (4.34) and (4.29), the BAM approach can easily be applied. For instance, Minnie (2002) discussed some aspects of the BAM model analytically. Matthaeus et al. (2003) compared the BAM theory for slab/2D composite geometry with test particle simulations and with the NLGC approach (see next Sect. 4.3). In their paper, it is shown that the BAM result is much smaller than the simulations. Furthermore, it is demonstrated by applying the more general nonlinear theory of Shalchi and Kourakis (see Chap. 2) that field line random walk in two-component turbulence behaves superdiffusively for the spectrum used in this book, which is the true reason for the singularity in the parameter κ_{2D}. For such nondiffusive regimes, the BAM formulation cannot be applied.

4.3 The Nonlinear Guiding Center Theory

A nonlinear theory for perpendicular transport has been proposed by Matthaeus et al. (2003), and is called the nonlinear guiding center (NLGC) theory. This approach can be considered as an improvement of the BAM model discussed in the previous section. Within the NLGC theory, *six models* or assumptions are used to derive a nonlinear integral equation for the perpendicular mean free path:

1. Again, the starting point of the NLGC approach is the assumption that *guiding centers follow magnetic field lines*

$$\tilde{v}_x = a v_z \frac{\delta B_x}{B_0}. \quad (4.36)$$

Actually, (4.36) describes guiding centers that follow field lines only for $a = 1$. Thus, the parameter a can be used to describe cases in which the guiding center motion deviates from the field line. In this work as well as in Matthaeus et al. (2003), we simply treat the parameter a as a proportionality constant that must be determined by comparing with simulations. Equation (4.36) can easily be combined with the TGK formulation to obtain

$$\begin{aligned} \kappa_{xx} &= \int_0^\infty \mathrm{d}t \ \langle \tilde{v}_x(t)\tilde{v}_x(0)\rangle, \\ &= \frac{a^2}{B_0^2}\int_0^\infty \mathrm{d}t \ \langle v_z(t)\delta B_x(t) v_z(0)\delta B_x^*(0)\rangle. \end{aligned} \quad (4.37)$$

2. To proceed, Matthaeus et al. assumed that the *fourth-order correlation can be replaced by a product of second-order correlations*

$$\kappa_{xx} = \frac{a^2}{B_0^2}\int_0^\infty \mathrm{d}t \ \langle v_z(t)v_z(0)\rangle \, \langle \delta B_x(t)\delta B_x^*(0)\rangle. \quad (4.38)$$

3. Next, *they modeled the parallel velocity correlation function by the (isotropic) exponential model*

$$V_{zz}(t) = \langle v_z(t)v_z(0)\rangle = \frac{v^2}{3}\,\mathrm{e}^{-vt/\lambda_\parallel}, \quad (4.39)$$

which is in accord with the model used in the BAM theory (see (4.24)). To proceed, the magnetic correlation function $R_{xx}(t) = \langle \delta B_x(t)\delta B_x^*(0)\rangle$ has to be replaced. As in QLT, the magnetic fields can be substituted by a Fourier representation

$$\delta B_x(\mathbf{x}, t) = \int \mathrm{d}^3k \ \delta B_x(\mathbf{k}, t)\,\mathrm{e}^{\mathrm{i}\mathbf{k}\cdot\mathbf{x}} \quad (4.40)$$

to find

$$R_{xx}(t) = \int \mathrm{d}^3k \ \left\langle \delta B_x(\mathbf{k}, t)\delta B_x^*(\mathbf{k}, 0)\,\mathrm{e}^{\mathrm{i}\mathbf{k}\cdot\Delta\mathbf{x}}\right\rangle, \quad (4.41)$$

where we again assumed homogeneous turbulence.

4. To proceed, *Corrsin's independence hypothesis* (Corrsin 1959; Salu and Montgomery 1977; McComb 1990) *has to be applied*

$$R_{xx}(t) = \int \mathrm{d}^3k \ P_{xx}(\mathbf{k}, t)\left\langle \mathrm{e}^{\mathrm{i}\mathbf{k}\cdot\mathbf{x}}\right\rangle. \quad (4.42)$$

5. Next, the characteristic function $\left\langle e^{i\mathbf{k}\cdot\mathbf{x}}\right\rangle$ must be approximated. Matthaeus et al. assumed a *Gaussian distribution of the particles* and, thus,

$$\left\langle e^{i\mathbf{k}\cdot\mathbf{x}}\right\rangle = e^{-\frac{1}{2}\langle(\Delta x)^2\rangle k_x^2-\frac{1}{2}\langle(\Delta y)^2\rangle k_y^2-\frac{1}{2}\langle(\Delta z)^2\rangle k_z^2}. \tag{4.43}$$

6. By additionally assuming that *parallel and perpendicular transport behave diffusively for all time scales* (e.g., $\langle(\Delta x)^2\rangle = 2t\kappa_{xx}$), the Gaussian solution of the three-dimensional diffusion equation (see, e.g., (1.93)) can be used:

$$\left\langle e^{i\mathbf{k}\cdot\mathbf{x}}\right\rangle = e^{-\kappa_{xx}k_x^2 t-\kappa_{yy}k_y^2 t-\kappa_{zz}k_z^2 t}. \tag{4.44}$$

This result can be combined with (4.42) and (4.38) to obtain

$$\kappa_{xx} = \frac{a^2}{B_0^2}\frac{v^2}{3}\int d^3k \int_0^\infty dt\; P_{xx}(\mathbf{k},t)\, e^{-vt/\lambda_\parallel-\kappa_{xx}k_x^2 t-\kappa_{yy}k_y^2 t-\kappa_{zz}k_z^2 t}. \tag{4.45}$$

By writing the correlation tensor as $P_{xx}(\mathbf{k},t) = \Gamma(\mathbf{k},t)P_{xx}(\mathbf{k})$, where we used the dynamical correlation function $\Gamma(\mathbf{k},t)$ and the static tensor $P_{xx}(\mathbf{k})$, and by assuming an exponential form for the time dependence,

$$\Gamma(\mathbf{k},t) = e^{-\gamma(\mathbf{k})t}, \tag{4.46}$$

we can solve the time-integral

$$\kappa_{xx} = \frac{a^2v^2}{3B_0^2}\int d^3k\, \frac{P_{xx}(\mathbf{k})}{v/\lambda_\parallel + \kappa_{xx}k_x^2 + \kappa_{yy}k_y^2 + \kappa_{zz}k_z^2 + \gamma(\mathbf{k})}. \tag{4.47}$$

This is the nonlinear integral equation of the NLGC theory (see Matthaeus et al. 2003, (7)). This integral equation can be used for turbulence models with purely magnetic fluctuations, where $\delta B_z = 0$, and for exponential forms of the dynamical correlation function.

Such forms can be found within the DT model and the NADT model (see Chap. 2). In Chap. 8, the latter dynamical turbulence model is combined with the NLGC theory to derive perpendicular mean free paths of charged particles propagating through the solar system. A combination of the NLGC theory and the different dynamical turbulence models can be found in Shalchi et al. (2004c).

4.4 Analytical Solutions of the NLGC Theory for Magnetostatic Slab Turbulence

Shalchi et al. (2004a) as well as Zank et al. (2004) explored the NLGC integral equation analytically for the magnetostatic model ($\gamma(\mathbf{k}) = 0$). In this case, and by assuming axisymmetric turbulence, (4.47) becomes

$$\kappa_\perp = \frac{a^2 v^2}{3B_0^2} \int d^3k \; \frac{P_{xx}(\mathbf{k})}{v/\lambda_\parallel + \kappa_\perp k_\perp^2 + \kappa_\parallel k_\parallel^2}. \tag{4.48}$$

To proceed, we restrict our calculations to slab geometry. In this case, (4.48) can be written as

$$\kappa_\perp = \frac{2a^2 v^2}{3B_0^2} C(\nu) l_{\mathrm{slab}} \delta B_{\mathrm{slab}}^2 \int_0^\infty dk_\parallel \; \frac{(1 + k_\parallel^2 l_{\mathrm{slab}}^2)^{-\nu}}{v/\lambda_\parallel + k_\parallel^2 \kappa_\parallel}, \tag{4.49}$$

where we again used (2.32) for the correlation tensor and (2.33) for the wave spectrum. By introducing the parameter

$$\varepsilon_{\mathrm{slab}} = \frac{3 l_{\mathrm{slab}}^2}{\lambda_\parallel^2} \tag{4.50}$$

and by applying the integral transformation $x = k_\parallel l_{\mathrm{slab}}$, we find for the perpendicular mean free path

$$\lambda_\perp = \frac{2a^2}{B_0^2} C(\nu) \frac{3\delta B_{\mathrm{slab}}^2 l_{\mathrm{slab}}^2}{\lambda_\parallel} \int_0^\infty dx \; \frac{\left(1 + x^2\right)^{-\nu}}{\varepsilon_{\mathrm{slab}} + x^2}. \tag{4.51}$$

The integral can be expressed by a hypergeometric function (see, e.g., Gradshteyn and Ryzhik 2000, Eq. (3.197.5))

$$\lambda_\perp = \frac{6a^2}{B_0^2} C(\nu) \frac{l_{\mathrm{slab}}^2 \delta B_{\mathrm{slab}}^2}{\lambda_\parallel} \frac{1}{2\varepsilon_{\mathrm{slab}}} B\left(\frac{1}{2}, \frac{1}{2} + \nu\right)$$
$$\times {}_2F_1\left(1, \frac{1}{2}, \nu + 1, \frac{\varepsilon_{\mathrm{slab}} - 1}{\varepsilon_{\mathrm{slab}}}\right), \tag{4.52}$$

where we used the Euler Beta function

$$B(x, y) \equiv \frac{\Gamma(x)\Gamma(y)}{\Gamma(x + y)}. \tag{4.53}$$

For small and large arguments, the hypergeometric function can be approximated (see, e.g., Abramowitz and Stegun 1974)

$${}_2F_1\left(1, \frac{1}{2}, \nu + 1, \frac{\varepsilon_{\mathrm{slab}} - 1}{\varepsilon_{\mathrm{slab}}}\right) \approx \begin{cases} \frac{2\nu}{2\nu - 1} & \text{for } \varepsilon_{\mathrm{slab}} \gg 1, \\ \sqrt{\pi} \frac{\Gamma(\nu+1)}{\Gamma(\nu+1/2)} \sqrt{\varepsilon_{\mathrm{slab}}} & \text{for } \varepsilon_{\mathrm{slab}} \ll 1. \end{cases} \tag{4.54}$$

By using these limits, we find for $\lambda_\parallel \gg \sqrt{3} l_{\text{slab}}$

$$\lambda_\perp \approx \sqrt{3}\pi a^2 C(\nu) l_{\text{slab}} \left(\frac{\delta B_{\text{slab}}}{B_0} \right)^2, \tag{4.55}$$

and for $\lambda_\parallel \ll \sqrt{3} l_{\text{slab}}$ we have

$$\frac{\lambda_\perp}{\lambda_\parallel} \approx \frac{a^2}{2} \frac{\delta B_{\text{slab}}^2}{B_0^2}. \tag{4.56}$$

For $a^2 = 1$, the latter formula corresponds to the results obtained within the NCA of Owens (1974). Equation (4.55) is similar to the quasilinear result. However, to recover the QLT-formula for $\lambda_\perp$, we have to choose $a^2 = \sqrt{3}/2$. Because the basic assumption of QLT is that guiding centers follow magnetic field lines ($a^2 = 1$), we cannot recover QLT from the NLGC theory.

As within the NCA and the BAM model, we find a diffusive behavior of perpendicular transport ($\langle (\Delta x)^2 \rangle \sim t$) in disagreement with test-particle simulations, where subdiffusion of the form $\langle (\Delta x)^2 \rangle \sim t^{1/2}$ has been discovered. It should be noted that, for slab geometry, the characteristic function is

$$\left\langle e^{i\mathbf{k}\cdot\mathbf{x}} \right\rangle = e^{-\kappa_{zz} k_z^2 t}. \tag{4.57}$$

Therefore, no assumptions about the perpendicular motion are required, nor has it been assumed that perpendicular transport behaves diffusively. Thus the reason for the failure of the NLGC theory for pure slab geometry must be found in the six steps described in the last paragraph. It is demonstrated in Chap. 7 that step 2 or step 3 must be the reason for the invalidity of this nonlinear transport theory.

4.5 NLGC Theory for Slab/2D Composite Geometry

As demonstrated, the NLGC theory is not valid for pure slab geometry. However, it is well known that diffusion is recovered for slab/2D composite geometry. In this case, the NLGC theory could become valid. In this paragraph, this possibility is explored by investigating the NLGC theory numerically. For the slab/2D composite model, analytical solutions of the integral equation have been derived in Shalchi et al. (2004a). For the two component model, the perpendicular mean free path can be written as (see (4.48))

$$\begin{aligned} \kappa_\perp = \frac{2a^2 v^2}{3B_0^2} C(\nu) \Bigg\{ & l_{\text{slab}} \delta B_{\text{slab}}^2 \int_0^\infty dk_\parallel \frac{(1 + k_\parallel^2 l_{\text{slab}}^2)^{-\nu}}{v/\lambda_\parallel + k_\parallel^2 \kappa_\parallel} \\ & + l_{2D} \delta B_{2D}^2 \int_0^\infty dk_\perp \frac{\left(1 + k_\perp^2 l_{2D}^2\right)^{-\nu}}{v/\lambda_\parallel + k_\perp^2 \kappa_\perp} \Bigg\}, \end{aligned} \tag{4.58}$$

where we used (2.50) and (2.51) to replace $P_{xx}^{2D}(\mathbf{k})$ in (4.47). By using the parameter

$$\varepsilon_{2D} = \frac{3 l_{2D}^2}{\lambda_\parallel \lambda_\perp} \tag{4.59}$$

and the integral transformation $x = k_\parallel l_{slab}$, we deduce

$$\lambda_\perp = \frac{2a^2}{B_0^2} C(\nu) \left\{ \frac{3\delta B_{slab}^2 l_{slab}^2}{\lambda_\parallel} \int_0^\infty \mathrm{d}x \, \frac{\left(1+x^2\right)^{-\nu}}{\varepsilon_{slab} + x^2} + \frac{3\delta B_{2D}^2 l_{2D}^2}{\lambda_\perp} \int_0^\infty \mathrm{d}x \, \frac{\left(1+x^2\right)^{-\nu}}{\varepsilon_{2D} + x^2} \right\} . \tag{4.60}$$

By solving the nonlinear integral equation numerically, we find the results shown in Fig. 4.1. By using the parallel mean free path from simulations and by choosing $a^2 = 1/3$, we can indeed achieve agreement with test-particle simulations. Within QLT, we find superdiffusion ($\lambda_\perp = \infty$) for the slab/2D composite model (see Chap. 3). The NCA has been derived only for slab geometry and can, therefore,

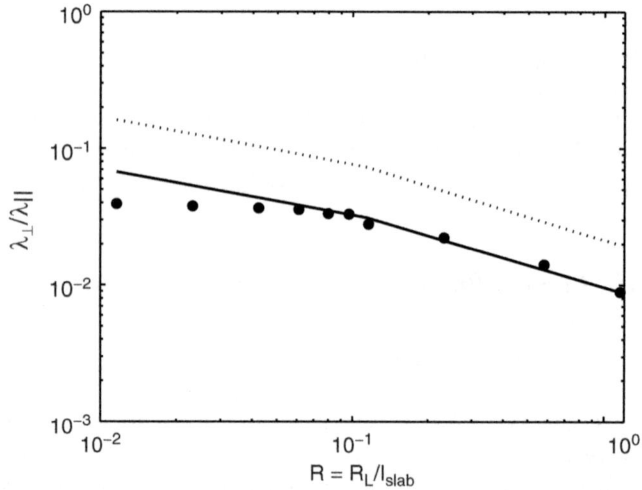

Fig. 4.1 Results of the NLGC theory for slab/2D composite geometry in comparison with test-particle simulations (*dots*). The NLGC theory results are obtained for two different values of a^2: $a^2 = 1$ corresponds to the *dotted line*, $a^2 = 1/3$ corresponds to the *solid line*, showing good agreement with simulations. Within QLT, we find superdiffusion ($\lambda_\perp = \infty$), which cannot be found in the numerical simulations. The NCA of Owens (1974) has been derived only for slab geometry and can, therefore, not be compared with the simulations. The application of the BAM theory is questionable due to the nondiffusive behavior of field line random walk in the composite model. Thus the BAM results are also not compared with the simulations and NLGC theory

not be compared with simulations performed for two-component turbulence. The BAM model has been compared with simulations in Matthaeus et al. (2003). The authors concluded that only the NLGC theory is in accord with simulations. It should be noted that the NLGC approach is the first theory that shows agreement with simulations. Furthermore, the NLGC theory will be used in Chap. 8 to reproduce heliospheric observations of $\lambda_{\perp}/\lambda_{\parallel}$. However, the nonlinear guiding center theory cannot describe subdiffusion in the slab model. Furthermore, we can only reproduce simulations by setting $a^2 = 1/3$. So far, there is no theoretical explanation available for this assumption. In the limit of small values of R ($R_{\mathrm{L}} \ll l_{\mathrm{slab}}$), we expect $a^2 = 1$ because, in this case, the guiding center assumption should be a good approximation. A theory for perpendicular as well as parallel transport can be achieved by combining the six steps of the NLGC theory with QLT. The resulting theory is called the weakly nonlinear theory and will be discussed in Chap. 5. An improved (or extended) NLGC theory that can also describe subdiffusion for slab geometry will be derived in Chap. 7.

Chapter 5
The Weakly Nonlinear Theory

The nonlinear theories discussed in Chap. 4 have been developed to solve the 90°-problem or the problem of perpendicular diffusion. As shown in Chap. 3, there is a third problem that occurs if the parallel mean free path is calculated linearly for nonslab models, namely the geometry problem. The test-particle simulations provide a much shorter parallel mean free path in comparison to the linear result. By combining the formulation of quasilinear theory with the six assumptions of the NLGC theory, a new transport theory can be derived that shows agreement with test-particle simulations for parallel as well as perpendicular diffusion. This new approach, which we refer to as weakly nonlinear theory, can solve the geometry problem and the problem of perpendicular transport by providing a coupled system of Fokker–Planck coefficients. Within the weakly nonlinear formulation, it can be demonstrated that pitch-angle and perpendicular diffusion cause resonance broadening. Mainly the resonance broadening due to the diffusive perpendicular motion makes the pitch-angle Fokker–Planck coefficient much larger and, thus, the parallel mean free path becomes smaller for most values of the rigidity.

5.1 The Basic Idea of a Nonlinear Transport Theory

Fokker–Planck coefficients can be calculated by using the TGK formulation (see e.g., (3.2) for purely magnetic fluctuations). To replace the guiding center velocity $\tilde{v}_i$, we can apply the assumption that guiding centers follow magnetic field lines (3.4), and to replace $\dot{\mu}$, we can use the z component of the equations of motion (3.3). Therewith, it can be demonstrated that the Fokker–Planck coefficients depend on fourth order correlation functions:

$$C_{ijlm}(\mathbf{x}, t) := \left\langle v_i(t) v_j(0) \delta B_l(\mathbf{x}(t)) \delta B_m^*(\mathbf{x}(0)) \right\rangle. \tag{5.1}$$

A. Shalchi, *Nonlinear Cosmic Ray Diffusion Theories*,
Astrophysics and Space Science Library,
DOI: 10.1007/978-3-642-00309-7_5, © Springer-Verlag Berlin Heidelberg 2009

With this definition, we have

$$\begin{aligned} D_{xx} &= \frac{1}{B_0^2}\mathfrak{Re}\int_0^\infty \mathrm{d}t\ C_{zzxx}(\mathbf{x},t), \\ D_{yy} &= \frac{1}{B_0^2}\mathfrak{Re}\int_0^\infty \mathrm{d}t\ C_{zzyy}(\mathbf{x},t), \\ D_{\mu\mu} &= \left(\frac{q}{cp}\right)^2 \mathfrak{Re}\int_0^\infty \mathrm{d}t \\ &\times \left[C_{xxyy}(\mathbf{x},t) - C_{xyyx}(\mathbf{x},t) - C_{yxxy}(\mathbf{x},t) + C_{yyxx}(\mathbf{x},t)\right]. \end{aligned} \tag{5.2}$$

To proceed, we replace the turbulent magnetic fields δB_i by Fourier transforms

$$\delta B_i(\mathbf{x},t) = \int \mathrm{d}^3k\ \delta B_i(\mathbf{k},t)\, \mathrm{e}^{\mathrm{i}\mathbf{k}\cdot\mathbf{x}}. \tag{5.3}$$

Hence,

$$C_{ijlm}(\mathbf{x},t) = \int \mathrm{d}^3k \int \mathrm{d}^3k' \left\langle v_i(t)v_j(0)\delta B_l(\mathbf{k},t)\delta B_m(\mathbf{k}',0)\,\mathrm{e}^{\mathrm{i}\left[\mathbf{k}\cdot\mathbf{x}(t)-\mathbf{k}'\cdot\mathbf{x}(0)\right]}\right\rangle. \tag{5.4}$$

A general calculation of this expression is difficult. Therefore, two approximations have to be applied:

1. Replace the fourth-order correlation by a product of two second-order correlations (Matthaeus et al. 2003)

$$\begin{aligned} C_{ijlm}(\mathbf{x},t) &\approx \left\langle v_i(t)v_j(0)\right\rangle \left\langle \delta B_l(\mathbf{x}(t))\delta B_m(\mathbf{x}(0))\right\rangle \\ &\equiv \int \mathrm{d}^3k \int \mathrm{d}^3k' \left\langle v_i(t)v_j(0)\right\rangle \left\langle \delta B_l(\mathbf{k},t)\delta B_m(\mathbf{k}',0)\,\mathrm{e}^{\mathrm{i}\left[\mathbf{k}\cdot\mathbf{x}(t)-\mathbf{k}'\cdot\mathbf{x}(0)\right]}\right\rangle. \end{aligned} \tag{5.5}$$

2. Apply Corrsin's independence hypothesis (Corrsin 1959; Salu and Montgomery 1977; McComb 1990; Matthaeus et al. 2003)

$$\begin{aligned} &\left\langle \delta B_l(\mathbf{k},t)\delta B_m(\mathbf{k}',0)\,\mathrm{e}^{\mathrm{i}\left[\mathbf{k}\cdot\mathbf{x}(t)-\mathbf{k}'\cdot\mathbf{x}(t)\right]}\right\rangle \\ &\qquad \approx \left\langle \delta B_l(\mathbf{k},t)\delta B_m(\mathbf{k}',0)\right\rangle \left\langle \mathrm{e}^{\mathrm{i}\left[\mathbf{k}\cdot\mathbf{x}(t)-\mathbf{k}'\cdot\mathbf{x}(t)\right]}\right\rangle. \end{aligned} \tag{5.6}$$

For homogeneous turbulence (see (2.4)), we find

$$C_{ijlm}(\mathbf{x},t) \approx \left\langle v_i(t)v_j(0)\right\rangle \int \mathrm{d}^3k\ P_{lm}(\mathbf{k},t)\left\langle \mathrm{e}^{\mathrm{i}\mathbf{k}\cdot\Delta\mathbf{x}(t)}\right\rangle, \tag{5.7}$$

with $\Delta\mathbf{x}(t) = \mathbf{x}(t) - \mathbf{x}(0)$. The characteristic function can be identified with the Fourier transform of the particle distribution function $f(\mathbf{x}, t)$

$$\left\langle \mathrm{e}^{\mathrm{i}\mathbf{k}\cdot\Delta\mathbf{x}(t)} \right\rangle = \int \mathrm{d}^3x \; f(\mathbf{x}, t)\, \mathrm{e}^{-\mathrm{i}\mathbf{k}\cdot\Delta\mathbf{x}(t)}. \tag{5.8}$$

A remaining problem is the determination of the following:

1. The velocity correlation functions $V_{ij}(t) = \left\langle v_i(t) v_j(0) \right\rangle$
2. The characteristic function $\Gamma(\mathbf{k}, t) = \left\langle \mathrm{e}^{\mathrm{i}\mathbf{k}\cdot\Delta\mathbf{x}(t)} \right\rangle$

If these two functions are known and if we assume knowledge of the tensor $P_{lm}(\mathbf{k}, t)$, the Fokker–Planck and diffusion coefficients can be calculated. In the following, the weakly nonlinear theory is discussed that uses simple models for the velocity correlation function and for the characteristic function similar to the models applied within the NLGC formulation.

5.2 The Weakly Nonlinear Resonance Function

First of all, we restrict our considerations to axisymmetric magnetostatic turbulence. Thereby, we have for the Fokker–Planck coefficients of perpendicular diffusion

$$D_{xx} = D_{yy} = \frac{1}{B_0^2} \mathfrak{Re} \int_0^\infty \mathrm{d}t \; C_{zzxx}(\mathbf{x}, t). \tag{5.9}$$

To determine the functions $C_{ijlm}(\mathbf{x}, t)$ that enter these equations as well as the corresponding formula for $D_{\mu\mu}$ (see (5.2)), we have to replace the velocity correlation functions and the characteristic function. Within the weakly nonlinear theory (WNLT), we apply the following models for the velocity correlation functions

$$\begin{aligned} V_{xx}(t) &= \langle v_x(t) v_x(0) \rangle = \langle v_x(t) v_x(0) \rangle_{\mathrm{QLT}}, \\ V_{yy}(t) &= \left\langle v_y(t) v_y(0) \right\rangle = \left\langle v_y(t) v_y(0) \right\rangle_{\mathrm{QLT}}, \\ V_{zz}(t) &= \langle v_z(t) v_z(0) \rangle = v^2 \mu^2 \, \mathrm{e}^{-\omega t}, \end{aligned} \tag{5.10}$$

with the pitch-angle scattering frequency

$$\omega = \frac{2 D_{\mu\mu}}{1 - \mu^2}. \tag{5.11}$$

The motivation for these models can easily be understood: the first two velocity correlation functions are important only for pitch-angle diffusion. We know (see e.g., Chap. 3) that QLT works very well for pitch-angle scattering in the slab model (except at 90°). Therefore, QLT should be a good approximation for the first two functions. The third velocity correlation function plays a crucial role if the perpendicular Fokker–Planck coefficient is calculated. Because QLT is always invalid for perpendicular transport independent of the turbulence model, the quasi-linear approximation is not appropriate to replace the parallel velocity correlation function $V_{zz}(t)$. Within the WNLT, we apply a simple model that describes a pitch-angle isotropization process, in agreement with Chap. 1.

If we model the v_z correlation function as described earlier, the unperturbed orbit is recovered for small times ($V_{zz}(t = 0) = v^2\mu^2$). For infinitely late times, we obtain $V_{zz}(t \to \infty) \to 0$ and the parallel velocity correlation function tends to zero. Physically, this model describes the pitch-angle isotropization process discussed in Chap. 1. In principle, a similar model could be used to model the v_x and v_y correlation functions. Here we still use unperturbed orbits to substitute these correlation functions. One reason for doing this is that, in WNLT, we consequently treat the particle's gyromotion as unperturbed. A second reason is that, in general, the v_x and v_y correlation functions can be dependent of pitch-angle diffusion, gyrophase diffusion, and perpendicular diffusion. Therefore, a realistic model for these correlation functions is difficult to formulate and would make the application of WNLT much more complicated.

Next, the characteristic function has to be estimated. If we assume a Gaussian distribution of the charged particles

$$f(\mathbf{x}, t) = \frac{1}{(2\pi)^{3/2}\sigma_x\sigma_y\sigma_z} \mathrm{e}^{-\frac{(x-\langle x\rangle)^2}{2\sigma_x^2}} \mathrm{e}^{-\frac{(y-\langle y\rangle)^2}{2\sigma_y^2}} \mathrm{e}^{-\frac{(z-\langle z\rangle)^2}{2\sigma_z^2}}, \tag{5.12}$$

we deduce for the characteristic function

$$\left\langle \mathrm{e}^{\mathrm{i}\mathbf{k}\cdot\Delta\mathbf{x}(t)} \right\rangle = \mathrm{e}^{-\frac{1}{2}\sum_{i=x,y,z}\sigma_i^2(t)k_i^2} \cdot \mathrm{e}^{\mathrm{i}\mathbf{k}\cdot\langle\Delta\mathbf{x}(t)\rangle}, \tag{5.13}$$

where we used the mean square displacements

$$\sigma_i^2 = \left\langle (\Delta x_i)^2 \right\rangle - \left(\langle \Delta x_i \rangle \right)^2. \tag{5.14}$$

Furthermore, we use the approximations

$$\begin{aligned} \sigma_x^2(t) &\approx 2tD_{xx}, \\ \sigma_y^2(t) &\approx 2tD_{yy}, \\ \sigma_z^2(t) &\approx 0, \\ \langle \Delta x_i \rangle &\approx \langle \Delta x_i \rangle_{\mathrm{QLT}} \equiv \Delta x_{i,\mathrm{QLT}} \end{aligned} \tag{5.15}$$

The first two approximations correspond to a late-time approximation. It is well known that, for early times, the particle motion is a ballistic motion and only a diffusive motion at late times (see Chap. 7). The first two approximations are accurate only if these small times are not important when calculating a diffusion coefficient. The third assumption neglects resonance broadening due to the partially statistical motion in the parallel direction. The motivation for this approach is that this effect is important only for the slab contribution to pitch-angle scattering. Because we know that QLT is accurate for the slab model (except at 90°), the resonance broadening effect due to $\sigma_z^2(t) \neq 0$ should be small. It should be noted, however, that this effect is essential if the 90°-problem should be solved (see Chap. 6). The fourth approximation assumes unperturbed orbits for the mean position of the particle. Because effects due to $\langle \Delta x_i \rangle \neq \langle \Delta x_i \rangle_{\text{QLT}}$ do not cause resonance broadening, we assume that this approximation is accurate. By combining these assumptions and models, we find

$$\begin{aligned}\left\langle e^{i\mathbf{k}\cdot\Delta\mathbf{x}(t)}\right\rangle &= e^{-\frac{t}{2}\sum_{i=x,y} D_{ii}k_i^2} \cdot e^{i\mathbf{k}\cdot\Delta\mathbf{x}_{\text{QLT}}(t)}, \\ &= e^{-\frac{1}{2}D_{xx}tk_\perp^2} \cdot e^{i\mathbf{k}\cdot\Delta\mathbf{x}_{\text{QLT}}(t)}, \end{aligned} \tag{5.16}$$

where we again used $D_{xx} = D_{yy}$. Therefore, we have

$$\begin{aligned} C_{ijlm}(\mathbf{x},t) &= v_{i,\text{QLT}}(t)v_{j,\text{QLT}}(0)\int d^3k\; P_{lm}(\mathbf{k},t)\, e^{-\frac{1}{2}D_{xx}tk_\perp^2} \cdot e^{i\mathbf{k}\cdot\Delta\mathbf{x}_{\text{QLT}}(t)}, \\ C_{zzxx}(\mathbf{x},t) &= v^2\mu^2\, e^{-\omega t}\int d^3k\; P_{xx}(\mathbf{k},t)\, e^{-\frac{1}{2}D_{xx}tk_\perp^2} \cdot e^{i\mathbf{k}\cdot\Delta\mathbf{x}_{\text{QLT}}(t)}, \end{aligned} \tag{5.17}$$

for $i,j,l,m = x,y$. These functions can be combined with (5.2) to determine the two Fokker–Planck coefficients $D_{\mu\mu}$ and $D_\perp$.

Within QLT, all Fokker–Planck coefficients are given by simple analytic forms (see Chap. 3). If these forms are compared with (5.17), the differences between QLT and WNLT are described by the exponential factors $e^{-\frac{1}{2}D_{xx}tk_\perp^2}$ and $e^{-\omega t}$. These factors change only the form of the resonance function. Within QLT, we have

$$R_n^{\text{QLT}}(\mathbf{k}) = \Re\mathfrak{e}\int_0^\infty dt\; e^{i(k_\parallel v_\parallel + n\Omega)t} = \pi\delta(k_\parallel v_\parallel + n\Omega) \tag{5.18}$$

for magnetostatic turbulence. By combining the exponential factors with the quasilinear resonance function, we obtain the WNLT from QLT by using the formal replacement:

$$R_n^{\text{QLT}}(\mathbf{k}) \to R_n^{\text{WNLT}}(\mathbf{k}), \tag{5.19}$$

where we introduced the weakly nonlinear resonance function

$$\begin{aligned} R_n^{\text{WNLT}}(\mathbf{k}) &= \Re\mathfrak{e}\int_0^\infty dt\; e^{i(k_\parallel v_\parallel + n\Omega)t} \cdot e^{-D_\perp k_\perp^2 t} \cdot e^{-\omega t} \\ &= \frac{D_\perp k_\perp^2 + \omega}{(D_\perp k_\perp^2 + \omega)^2 + (k_\parallel v_\parallel + n\Omega)^2}, \end{aligned} \tag{5.20}$$

with

$$\omega = \begin{cases} \frac{2D_{\mu\mu}}{1-\mu^2} & \text{for perpendicular diffusion} \\ 0 & \text{for pitch-angle diffusion.} \end{cases} \tag{5.21}$$

Equations (5.20) and (5.21) can be derived by combining (5.2) and (5.9) with (5.17). Obviously, we find a Breit–Wigner type resonance function for R_n^{WNLT}. In WNLT, resonance broadening is caused by perpendicular and pitch-angle diffusion. QLT can easily be recovered if we set $D_{\mu\mu} = 0$ and $D_\perp = 0$ on the right-hand side of (5.20).

5.3 The Nonlinear Fokker–Planck Coefficients for Two-component Turbulence

It is a simple matter to derive weakly nonlinear Fokker–Planck coefficients, because we only have to replace the resonance functions in the corresponding QLT formulas for $D_{\mu\mu}$ and $D_\perp$. For instance, we can use (3.22) to describe weakly nonlinear pitch-angle diffusion by replacing R_n herein with (5.20). In the following, we apply the slab/2D composite model to determine the weakly nonlinear Fokker–Planck coefficients.

5.3.1 The Fokker–Planck Coefficient $D_{\mu\mu}^{\mathrm{slab}}$

For pitch-angle diffusion in the magnetostatic slab model, we have $k_\perp = 0$ and $\omega = 0$ in (5.20) and the resonance function is equal to the delta function of QLT

$$R_n^{\mathrm{slab}}(\mathbf{k}) = \pi\delta(k_\parallel v_\parallel + n\Omega) \tag{5.22}$$

and we can use the QLT-formula for $D_{\mu\mu}$ (see (3.34)):

$$D_{\mu\mu}^{\mathrm{slab}} = \frac{2\pi^2\Omega^2(1-\mu^2)}{B_0^2}\int_0^\infty dk_\parallel\, g^{\mathrm{slab}}(k_\parallel)\left[\delta(k_\parallel v_\parallel + \Omega) + \delta(k_\parallel v_\parallel - \Omega)\right]. \tag{5.23}$$

For convenience, we define the dimensionless parameters

$$R = \frac{R_{\mathrm{L}}}{l_{\mathrm{slab}}},$$
$$\xi = \frac{l_{\mathrm{slab}}}{l_{\mathrm{2D}}},$$

$$\tilde{\omega} = \frac{l_{\text{slab}}}{v}\omega,$$
$$\tilde{D}_{\mu\mu} = \frac{l_{\text{slab}}}{v}D_{\mu\mu},$$
$$\tilde{D}_{\perp} = \frac{1}{v l_{\text{slab}}}D_{\perp}, \tag{5.24}$$

with $R_{\text{L}} = v/\Omega$. With these parameters and with the spectrum from (2.33), and by solving the integral over $k_{\parallel}$, we find

$$\tilde{D}_{\mu\mu}^{\text{slab}} = \pi C(\nu)(1-\mu^2) \mid \mu \mid^{2\nu-1} R^{2\nu-2}\frac{\delta B_{\text{slab}}^2}{B_0^2}\left(\mu^2 R^2 + 1\right)^{-\nu}. \tag{5.25}$$

5.3.2 The Fokker–Planck Coefficient $D_{\mu\mu}^{2D}$

For pitch-angle diffusion in pure 2D geometry, we have $\omega = 0$ and $k_{\parallel} = 0$. Thus, the resonance function from (5.20) becomes

$$R_n^{2D}(\mathbf{k}) = \frac{D_{\perp}k_{\perp}^2}{(D_{\perp}k_{\perp}^2)^2 + (n\Omega)^2}. \tag{5.26}$$

To calculate $D_{\mu\mu}^{2D}$ within WNLT, we can start from (3.52):

$$D_{\mu\mu}^{2D} = \frac{2\pi\Omega^2(1-\mu^2)}{B_0^2}\int_0^{\infty} dk_{\perp}\, g^{2D}(k_{\perp}) \sum_{n=-\infty}^{\infty} R_n^{2D}(\mathbf{k})\frac{n^2 J_n^2(W)}{W^2} \tag{5.27}$$

with $W = k_{\perp}R_{\text{L}}\sqrt{1-\mu^2}$. In combination with (5.26), this can be written as

$$D_{\mu\mu}^{2D} = \frac{4\pi(1-\mu^2)}{B_0^2}\int_0^{\infty} dk_{\perp}\, g^{2D}(k_{\perp})D_{\perp}k_{\perp}^2 \sum_{n=1}^{\infty}\frac{n^2}{n^2 + \left(D_{\perp}k_{\perp}^2/\Omega\right)^2}\frac{J_n^2(W)}{W^2}. \tag{5.28}$$

With the approximation of Shalchi and Schlickeiser (2004a, (19))

$$\sum_{n=1}^{\infty}\frac{n^2}{n^2+y^2}J_n^2(x) \approx \frac{x^2}{2}\frac{1}{x^2+2y^2+2}, \tag{5.29}$$

we find

$$D_{\mu\mu}^{2D} = \frac{2\pi\Omega^2(1-\mu^2)}{B_0^2}\int_0^{\infty} dk_{\perp}\, g^{2D}(k_{\perp})\frac{D_{\perp}k_{\perp}^2}{W^2\Omega^2 + 2(D_{\perp}k_{\perp}^2)^2 + 2\Omega^2}. \tag{5.30}$$

By using the spectrum from (2.51), the parameters of (5.24), and the integral transformation $x = k_{\perp} l_{2D}$, we find

$$\tilde{D}_{\mu\mu}^{2D} = 2C(\nu)\frac{1-\mu^2}{R^2}\frac{\delta B_{2D}^2}{B_0^2}\int_0^{\infty} dx\ (1+x^2)^{-\nu} \times \frac{\tilde{D}_{\perp}\xi^2 x^2}{(1-\mu^2)\xi^2 x^2/2 + \left[\tilde{D}_{\perp}\xi^2 x^2\right]^2 + R^{-2}}. \tag{5.31}$$

The QLT result for this Fokker–Planck coefficient can be recovered for $\tilde{D}_{\perp} \to 0$. In this limit, we find the expected result, $\tilde{D}_{\mu\mu}^{2D} = 0$, in agreement with Chap. 3.

5.3.3 The Fokker–Planck Coefficient $D_{\perp}^{\text{slab}}$

For slab geometry and perpendicular diffusion, the weakly nonlinear resonance function (see (5.20)) becomes

$$R_n(\mathbf{k}) = \frac{\omega}{\omega^2 + (k_{\parallel} v_{\parallel} + n\Omega)^2}. \tag{5.32}$$

Combined with (3.29) and by replacing $P_{ij}(\mathbf{k})$ herein with (2.32), we find for the Fokker–Planck coefficient of perpendicular diffusion

$$D_{\perp}^{\text{slab}} = \frac{4\pi v^2 \mu^2}{B_0^2}\int_0^{\infty} dk_{\parallel}\ g^{\text{slab}}(k_{\parallel})\frac{\omega}{\omega^2 + k_{\parallel}^2 v_{\parallel}^2}. \tag{5.33}$$

With the spectrum from (2.33), the parameters of (5.24), and the integral transformation $x = k_{\parallel} l_{\text{slab}}$, this can be rewritten as

$$\tilde{D}_{\perp}^{\text{slab}} = 2C(\nu)\mu^2\frac{\delta B_{\text{slab}}^2}{B_0^2}\int_0^{\infty} dx\ (1+x^2)^{-\nu}\frac{\tilde{\omega}}{\tilde{\omega}^2 + \mu^2 x^2}. \tag{5.34}$$

By investigating the limit $\omega \to 0$ (corresponding to $\tilde{\omega} \to 0$), we can easily derive the quasilinear result (3.47) as expected.

5.3.4 The Fokker–Planck Coefficient $D_{\perp}^{2D}$

For pure 2D geometry and perpendicular diffusion, the weakly nonlinear resonance function is

$$R_n(\mathbf{k}) = \frac{D_{\perp}k_{\perp}^2 + \omega}{(D_{\perp}k_{\perp}^2 + \omega)^2 + (n\Omega)^2}. \tag{5.35}$$

For the calculation of $D_{\perp}^{2D}$, we start with (3.29). By using (2.50) for $P_{lm}(\mathbf{k})$ herein, the Fokker–Planck coefficient of perpendicular diffusion becomes

$$D_{\perp}^{2D} = \frac{\pi v^2 \mu^2}{\Omega B_0^2} \int_0^{\infty} dk_{\perp}\, g^{2D}(k_{\perp}) V H(V, W), \tag{5.36}$$

where we used

$$H(V, W) = \sum_{n=-\infty}^{\infty} \frac{J_n^2(W)}{V^2 + n^2} \tag{5.37}$$

and

$$\begin{aligned} V &= (D_{\perp} k_{\perp}^2 + \omega)/\Omega, \\ W &= R_L k_{\perp} \sqrt{1-\mu^2}. \end{aligned} \tag{5.38}$$

As demonstrated in Shalchi and Schlickeiser (2004b), the function H can be approximated for different cases:

$$\begin{aligned} H(V \ll 1, W \gg 1) &\approx \frac{J_0^2(W)}{V^2} \approx \frac{2}{\pi W V^2} \cos^2(W - \frac{\pi}{4}) \approx \frac{1}{\pi W V^2}, \\ H(V \ll 1, W \ll 1) &\approx \frac{1}{V^2}, \\ H(V \gg 1, V \ll W) &\approx \frac{1}{V\, W}, \\ H(V \gg 1, V \gg W) &\approx \frac{1}{V^2}. \end{aligned} \tag{5.39}$$

Using the integral transformation $x = k_{\perp} l_{2D}$, the parameters of (5.24), and the spectrum of (2.51), we obtain

$$\tilde{D}_{\perp}^{2D} = 2C(\nu)\mu^2 R \frac{\delta B_{2D}^2}{B_0^2} \int_0^{\infty} dx\, (1+x^2)^{-\nu}\, \tilde{V}\, H(\tilde{V}, \tilde{W}), \tag{5.40}$$

with

$$\begin{aligned} \tilde{V} &= R \tilde{D}_{\perp} \xi^2 x^2 + R \tilde{\omega}, \\ \tilde{W} &= R \xi x \sqrt{1-\mu^2}. \end{aligned} \tag{5.41}$$

The QLT result for this Fokker–Planck coefficient is $\tilde{D}_{\perp}^{2D} \to \infty$.

5.4 Results of WNLT for the Parallel and the Perpendicular Mean Free Path

Within WNLT, we find a coupled system for the Fokker–Planck coefficients $\tilde{D}_{\mu\mu}^{\text{slab}}$ (5.25), $\tilde{D}_{\mu\mu}^{2\text{D}}$ (5.31), $\tilde{D}_{\perp}^{\text{slab}}$ (5.34), and $\tilde{D}_{\perp}^{2\text{D}}$ (5.40). By applying

$$\begin{aligned}\tilde{D}_{\mu\mu} &= \tilde{D}_{\mu\mu}^{\text{slab}} + \tilde{D}_{\mu\mu}^{2\text{D}},\\ \tilde{D}_{\perp} &= \tilde{D}_{\perp}^{\text{slab}} + \tilde{D}_{\perp}^{2\text{D}},\end{aligned} \tag{5.42}$$

the pitch-angle Fokker–Planck coefficients for slab/2D composite geometry can easily be computed numerically. With these results, we can determine the mean free paths by simply integrating over the pitch-angle using (see (3.1))

$$\lambda_{\parallel} = \frac{3}{4} l_{\text{slab}} \int_0^1 \mathrm{d}\mu \, \frac{(1-\mu^2)^2}{\tilde{D}_{\mu\mu}}, \tag{5.43}$$

$$\lambda_{\perp} = 3 l_{\text{slab}} \int_0^1 \mathrm{d}\mu \, \tilde{D}_{\perp}. \tag{5.44}$$

For the numerical evaluation of (5.43) and (5.44), we use the following set of parameters to compare our theory with simulations:

$$\begin{aligned}\delta B &= B_0,\\ \delta B_{\text{slab}}^2 &= 0.2\,\delta B^2,\\ \delta B_{2\text{D}}^2 &= 0.8\,\delta B^2,\\ \nu &= 5/6,\\ l_{\text{slab}} &= 0.030\,\text{AU},\\ l_{2\text{D}} &= 0.1 l_{\text{slab}} = 0.003\,\text{AU}.\end{aligned} \tag{5.45}$$

In the next paragraph, some figures also show results for different parameters like pure slab geometry. In this case, the parameters that are different from (5.45) are explicitly declared. In the following, several results obtained within WNLT are shown. More results can be found in Shalchi et al. (2004b).

5.4.1 *The Nonlinear Fokker–Planck Coefficients $D_{\mu\mu}$ and $D_{\perp}$*

Figure 5.1 shows the results for the pitch-angle Fokker–Planck coefficient $D_{\mu\mu}^{\text{NL}}$. All pitch-angle Fokker–Planck coefficients have the same form, which is a surprising result. Normally, the form of the pitch-angle Fokker–Planck coefficient results from gyroresonance. It seems that WNLT provides us with similar results even in pure 2D geometry where we do not have gyroresonance. The QLT result for composite geometry is exactly by a factor 5 smaller than the QLT result for pure slab geometry,

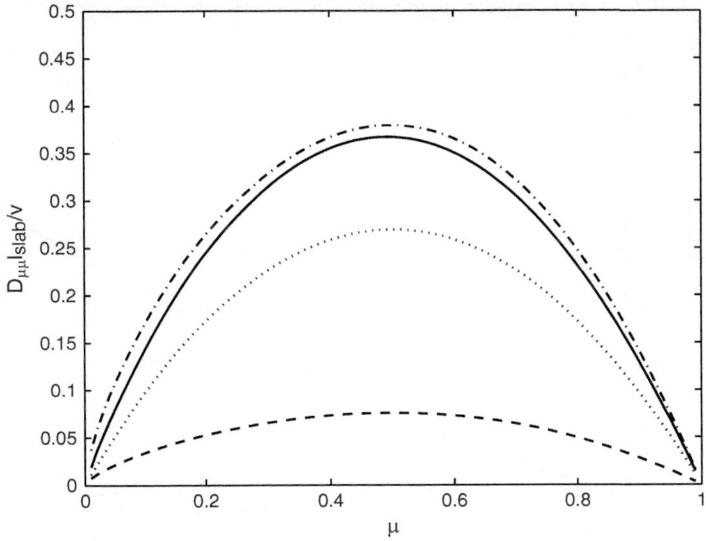

Fig. 5.1 The nonlinear pitch-angle Fokker–Planck coefficient for composite geometry (*solid line*) and pure 2D (*dotted line*) in comparison with quasilinear Fokker–Planck coefficients for composite geometry (*dashed line*) and pure slab (*dash-dotted line*). For this plot, $R = R_L/l_{slab} = 0.1$ and $l_{slab} = l_{2D} = 0.03$ AU have been used

whereas the WNLT result for composite geometry is nearly equal to the quasilinear slab result. Figure 5.2 shows the pitch-angle dependence of the Fokker–Planck coefficient of perpendicular diffusion $D_{\perp}^{NL}$ in comparison with the slab QLT result. The QLT result for composite geometry is $D_{\perp} \to \infty$ (superdiffusion).

5.4.2 $\lambda_{\parallel}$, $\lambda_{\perp}$, and $\lambda_{\perp}/\lambda_{\parallel}$ for Two-component Turbulence

Figure 5.3 shows simulations of the parallel mean free path (Qin 2002; Qin et al. 2002a, b), magnetostatic QLT-results for pure slab, QLT-results for 20% slab turbulence, and the WNLT results for 20% slab turbulence. As demonstrated, the WNLT agrees much better with simulations than QLT. The rigidity dependence for small and medium rigidities is approximately $\lambda_{\parallel} \sim R^{0.6}$, which is in contrast to QLT-results ($\lambda_{\parallel} \sim R^{1/3}$) but in agreement with simulations (see Minnie 2002). For high rigidities ($R \gg 1$), QLT seems to be recovered. Our numerical calculations also show that the reason for the difference between QLT and simulations arises from the perpendicular diffusion of the particles. Therefore, we come to the conclusion that perpendicular diffusion has a strong influence on the parallel mean free path. In Fig. 5.4, the new results for the perpendicular mean free path are illustrated. Shown are the results of WNLT (solid line) in comparison to simulations (dots) and results of the NLGC theory. To calculate $\lambda_{\perp}$ with the NLGC theory, we need the parallel mean free path as an input. Here, we use the parallel mean free path from simulations

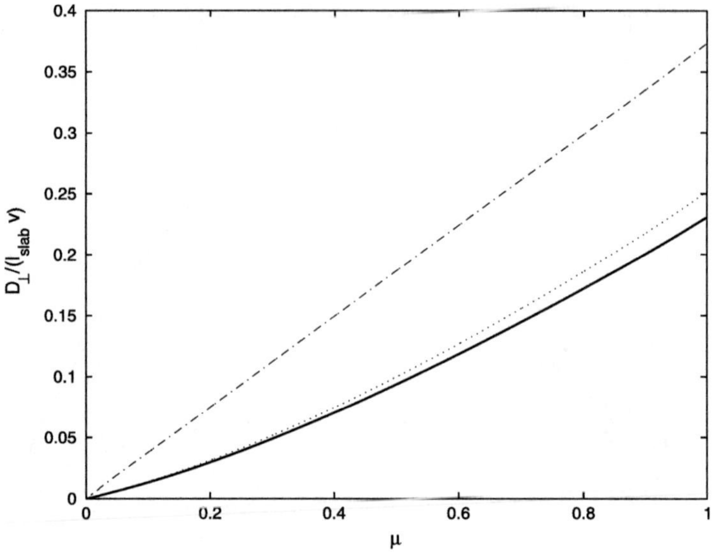

Fig. 5.2 The nonlinear Fokker–Planck coefficient of perpendicular diffusion for composite geometry (*solid line*) and for pure 2D (*dotted line*) in comparison with the QLT slab result (*dash-dotted line*). For this plot, $R = R_L / l_{slab} = 0.1$ and $l_{slab} = l_{2D} = 0.03$ AU have been used

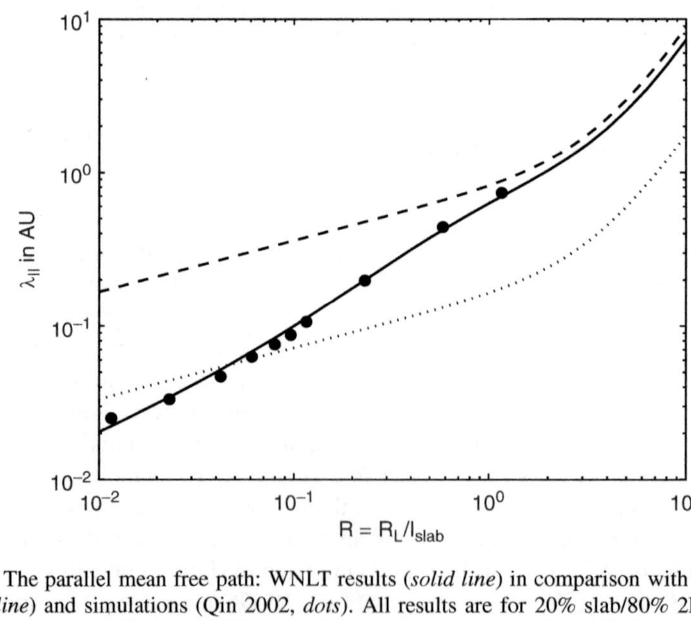

Fig. 5.3 The parallel mean free path: WNLT results (*solid line*) in comparison with QLT results (*dashed line*) and simulations (Qin 2002, *dots*). All results are for 20% slab/80% 2D geometry. Also shown are the well-known QLT results for pure slab geometry (*dotted line*)

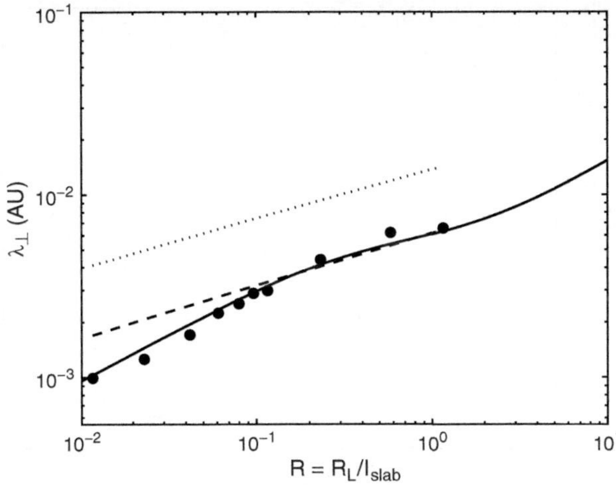

Fig. 5.4 The perpendicular mean free path: WNLT results (*solid line*) in comparison with NLGC results for $a^2 = 1$ (*dotted line*) and $a^2 = 1/3$ (*dashed line*), and simulations (Qin 2002, *dots*). All results are for 20% slab/80% 2D geometry

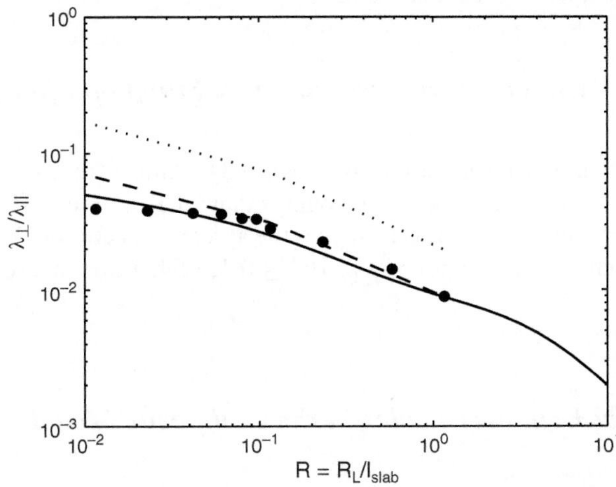

Fig. 5.5 The ratio $\lambda_\perp/\lambda_\parallel$: WNLT results (*solid line*) in comparison with NLGC results $a^2 = 1$ (*dotted line*) and $a^2 = 1/3$ (*dashed line*), and simulations (Qin 2002, *dots*). All results are for 20% slab/80% 2D geometry

(see Fig. 5.3). The NLGC theory contains a fitting parameter a. We compare our results with NLGC results for two different values of a: the value $a = 1$ (dotted line) should be correct in the weak turbulence limit, whereas the value $a = 1/\sqrt{3}$ (dashed line) provides the best agreement with simulations. For completeness, also the ratio $\lambda_\perp/\lambda_\parallel$ is shown (Fig. 5.5). The results shown are quite similar to the NLGC theory results if the parameter a is assumed to be $a^2 = 1/3$.

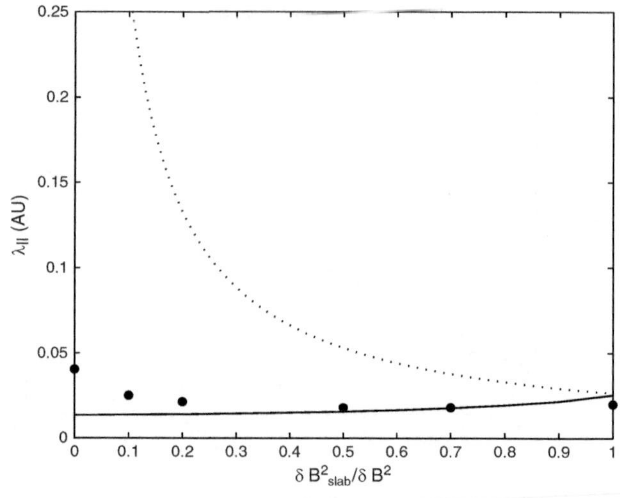

Fig. 5.6 The parallel mean free path as a function of the ratio $\delta B^2_{\text{slab}}/\delta B^2$ for moderate amplitude turbulence ($\delta B/B_0 = 1$). Shown are results of the weakly nonlinear theory (*solid line*), simulations (*dots*), and QLT (*dotted line*). For pure 2D ($\delta B^2_{\text{slab}}/\delta B^2 = 0$), the QLT prediction is $\lambda_{\parallel} = \infty$. To obtain these results, $R = 0.005$ has been used

5.4.3 The Parallel Mean Free Path as a Function of $\delta B^2_{\text{slab}}/\delta B^2$

In Fig. 5.6, the parallel mean free path is shown as a function of the ratio $\delta B^2_{\text{slab}}/\delta B^2 = \delta B^2_{\text{slab}}/(\delta B^2_{\text{slab}} + \delta B^2_{\text{2D}})$. It is easy to recognize that QLT disagrees with simulations for nonslab turbulence geometries, whereas the agreement between WNLT and simulations is much better. Only for $\delta B^2_{\text{slab}}/\delta B^2 \leq 0.1$, WNLT does not seem to be very accurate.

5.4.4 Equal Bendover Scales in the Composite Model

One of the key input parameters for theoretical calculations is the ratio of the bendover scales $l_{\text{2D}}/l_{\text{slab}}$. There are indications from laboratory experiments such as Robinson and Rusbridge (1971) that the parallel correlation scale may be one magnitude larger than the perpendicular scale in an evolving magnetofluid plasma. In astrophysical systems, this ratio is unclear, but in previous articles (e.g., Matthaeus et al. 2003), the assumption $l_{\text{2D}}/l_{\text{slab}} = 0.1$ was used. In Fig. 5.7, simulations and WNLT are compared with each other for equal bendover scales ($l_{\text{2D}} = l_{\text{slab}}$).

As demonstrated in Fig. 5.7, the deviation between QLT and simulations is smaller for smaller $l_{\text{slab}}/l_{\text{2D}}$. For $l_{\text{2D}}/l_{\text{slab}} = 0.1$, for instance, we have $\lambda_{\parallel}^{\text{QLT}}/\lambda_{\parallel}^{\text{Sim}}$ $(R = 0.01) \approx 9$ and, for $l_{\text{2D}}/l_{\text{slab}} = 1$, we find $\lambda_{\parallel}^{\text{QLT}}/\lambda_{\parallel}^{\text{Sim}}(R = 0.01) \approx 4$. This

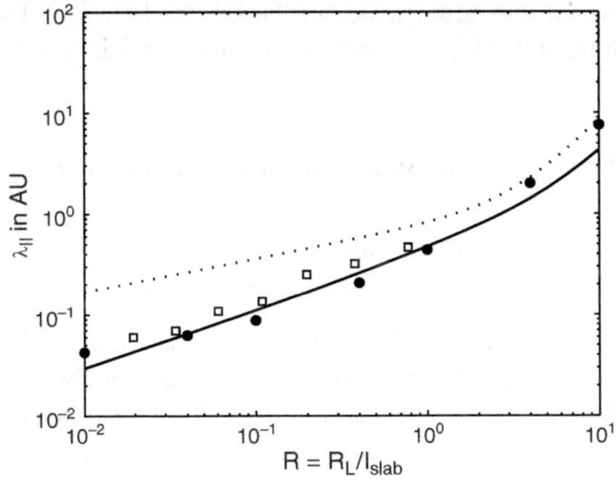

Fig. 5.7 $\lambda_{\parallel}$ for equal bendover scales ($l_{2D} = l_{slab}$) and composite geometry. The simulations obtained by applying the test-particle code developed by Qin et al. (2002a, b, see *dots*) are compared with the simulations of Giacalone and Jokipii (1999, *squares*), with QLT (*dotted line*) and with WNLT (*solid line*)

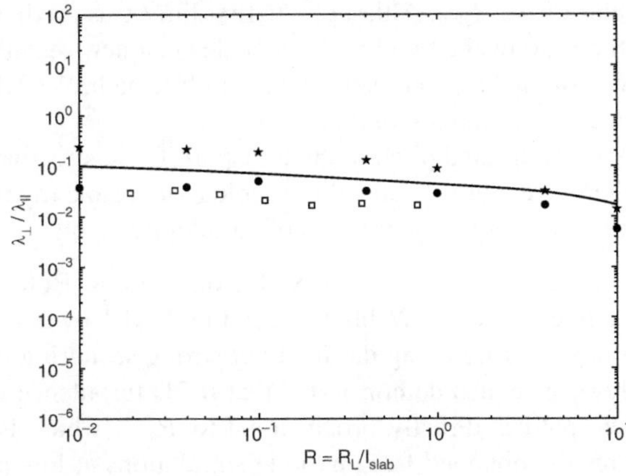

Fig. 5.8 $\lambda_{\perp}/\lambda_{\parallel}$ for equal bendover scales ($l_{2D} = l_{slab}$) and composite geometry. The simulations obtained by applying the test-particle code developed by Qin et al. (2002a, b, see *dots*) are compared with the simulations of Giacalone and Jokipii (1999, *squares*), with NLGC-theory (*stars*), and with WNLT (*solid line*). The QLT result is again $\lambda_{\perp}/\lambda_{\parallel} = \infty$

effect could be important to justify quasilinear results that are used in Chap. 8 to reproduce observations in the solar wind. In Fig. 5.8, the ratio $\lambda_{\perp}/\lambda_{\parallel}$ is shown for equal scales. Except for large rigidities, the values for equal scales are similar in comparison to the results obtained for $l_{2D} = 0.1 l_{slab}$. In both figures, also the simulations of Giacalone and Jokipii (1999) are shown. Their simulations are already compared with the simulations performed by Qin (2002) in Tautz et al. (2006a). There, it was stated that the discovered difference is caused by the different bendover scales ($l_{2D}/l_{slab} = 1$ in the simulations of Giacalone and Jokipii (1999) and $l_{2D}/l_{slab} = 0.1$ in the Qin simulations). As shown in Figs. 5.7 and 5.8, this statement is true: if we also choose equal scales in the Qin code, we find similar results as found by Giacalone and Jokipii for parallel and perpendicular diffusion.

5.5 Is the Weakly Nonlinear Theory Reasonable?

In the previous section of this chapter, some aspects of the WNLT are discussed. The most important conclusions obtained in this chapter are

- Perpendicular and pitch-angle diffusion lead to resonance broadening. By assuming diffusive behavior of perpendicular transport and by using an exponential model for the parallel velocity correlation function, we can enforce a Breit–Wigner form of the resonance function. As shown, such resonance broadening effects are in general important and cannot be neglected.
- One of the advantages of WNLT is that previous QLT results can be used. By the formal replacement of resonance functions, WNLT formulas for Fokker–Planck coefficients can be derived.
- The nonlinear effect can be suppressed by choosing $\delta B/B_0 \to 0$ or by assuming larger values of l_{2D}/l_{slab}. Although the first limit was predictable, the strong influence of the ratio of the two bendover scales is a new result. Because of the strong influence of l_{2D}/l_{slab} on scattering mean free path, WNLT also motivates an observational determination of this ratio.
- For pure slab geometry and pitch-angle transport, QLT is recovered. Obviously, WNLT can, therefore, not solve the 90°-problem of cosmic ray transport.
- WNLT can solve the geometry problem of parallel transport.

le Roux and Webb (2007) derived the WNLT from a BGK–Boltzmann approach. From this approach le Roux and Webb deduced the WNLT by assuming that particles undergo weak scattering along the field but strong scattering transverse to the field. These authors have also demonstrated that if 2D turbulence is dominant, the parallel mean free path is directly proportional to $R^{2/3}$, which is approximately the same as the results obtained from particle simulations at low particle energies (see e.g., Fig. 5.3).

It will be demonstrated in Chap. 8 that the geometry effect described by WNLT is important for reproducing the measured abundance ratio of primary-to-secondary cosmic ray nuclei. Thus, the WNLT is also useful for astrophysical applications and not only for reproducing computer simulations.

Although WNLT can be considered as an advanced transport theory that can describe parallel and perpendicular transport in agreement with simulations, there are still some problems left associated with the approximations used in WNLT. For instance, WNLT cannot be applied for cases where 90°-scattering is essential. By applying a second-order approach, it will be demonstrated in the next chapter how this problem can be investigated. A further problem will be discussed in Chap. 7: both WNLT and NLGC theory cannot correctly describe subdiffusive perpendicular transport in pure slab models.

Chapter 6
The Second-order QLT

As shown in the previous chapter, the weakly nonlinear theory cannot solve the 90°-problem of cosmic ray transport. Furthermore, previous theories that have been developed to describe pitch-angle diffusion at 90° are not reliable or at least intractable. In this section, an alternative approach is described, which is based on quasilinear theory. By calculating corrections of the unperturbed orbit and by combining the improved particle trajectories with equations for the pitch-angle diffusion coefficient, a second-order theory can be formulated. As demonstrated here, the second-order QLT can describe pitch-angle diffusion for all values of μ in agreement with test-particle simulations. As an example, particle propagation in turbulent systems with steep wave spectra is considered.

6.1 Nonlinear Pitch-angle Diffusion in Pure Slab Turbulence

As demonstrated in the previous chapter, the key input in a nonlinear transport theory is the fourth-order correlation function (see (5.1)). For magnetostatic slab turbulence, by assuming Gaussian statistics for the particle distribution and by approximating the fourth-order correlation by a product of two second-order correlations, we find, according to (5.7)

$$\begin{aligned} C_{ijlm}(t) &= \left\langle v_i(t) v_j(0) \right\rangle \int d^3k \; P_{lm}(\mathbf{k}) \left\langle e^{i\mathbf{k}\cdot\Delta\mathbf{x}} \right\rangle, \\ &= 4\pi \left\langle v_i(t) v_j(0) \right\rangle \int_0^\infty dk_\parallel \; g^{\text{slab}}(k_\parallel) \left(\delta_{lm} - \frac{k_l k_m}{k^2} \right) \Gamma(k_\parallel, t), \quad (6.1) \end{aligned}$$

with the characteristic function

$$\Gamma(k_\parallel, t) \equiv \left\langle e^{ik_\parallel \Delta z(t)} \right\rangle = e^{ik_\parallel \langle \Delta z(t) \rangle} \, e^{-\frac{1}{2}\sigma_z^2(t) k_\parallel^2}. \quad (6.2)$$

A. Shalchi, *Nonlinear Cosmic Ray Diffusion Theories*,
Astrophysics and Space Science Library,
DOI: 10.1007/978-3-642-00309-7_6,

For the pitch-angle Fokker–Planck coefficient, we then find, by combining (6.1) with (5.2),

$$\begin{aligned} D_{\mu\mu} &= \left(\frac{q}{cp}\right)^2 \int_0^\infty \mathrm{d}t \, \left[C_{xxyy}(t) + C_{yyxx}(t)\right], \\ &= 8\pi \left(\frac{\Omega}{vB_0}\right)^2 \int_0^\infty \mathrm{d}k_\parallel \, g^{\mathrm{slab}}(k_\parallel) \sum_{i=x,y} \int_0^\infty \mathrm{d}t \, \langle v_i(t) v_i(0)\rangle \, \Gamma(k_\parallel, t). \end{aligned} \tag{6.3}$$

The basic idea of second-order quasilinear theory (SOQLT) of Shalchi (2005a) is to determine the parameters $V_{ij}(t) = \langle v_i(t)v_j(0)\rangle$ and $\Gamma(k_\parallel, t)$ by using the usual (first-order) QLT. The calculated parameters can be seen as improved orbits that can be used to substitute the corresponding parameters in (6.3). The resulting Fokker–Planck coefficient can be called a second-order coefficient. During this chapter, the following notation is used: for example, $D_{\mu\mu}^{(1)}$ means that $D_{\mu\mu}$ is calculated by using QLT. $D_{\mu\mu}^{(2)}$ indicates that improved (linearly calculated) orbits are used to determine $D_{\mu\mu}$. The second-order pitch-angle Fokker–Planck coefficient can, therefore, be written as

$$D_{\mu\mu}^{(2)} = 8\pi \left(\frac{\Omega}{vB_0}\right)^2 \int_0^\infty \mathrm{d}k_\parallel \, g^{\mathrm{slab}}(k_\parallel) \sum_{i=x,y} \int_0^\infty \mathrm{d}t \, V_{ii}^{(1)}(t) \Gamma^{(1)}(k_\parallel, t). \tag{6.4}$$

In the following, the first-order functions $V_{ii}^{(1)}(t)$ and $\Gamma^{(1)}(\mathbf{k}, t)$ are calculated using QLT.

6.1.1 The Quasilinear Velocity Correlation Function

The equations of motion (1.6) can be reformulated as integral equations:

$$\begin{aligned} v_x(\tau) &= v_x(0) + \Omega \int_0^\tau \mathrm{d}t \, \left[v_y(t) - v_z(t)\frac{\delta B_y(t)}{B_0}\right], \\ v_y(\tau) &= v_y(0) + \Omega \int_0^\tau \mathrm{d}t \, \left[-v_x(t) + v_z(t)\frac{\delta B_x(t)}{B_0}\right], \\ v_z(\tau) &= v_z(0) + \Omega \int_0^\tau \mathrm{d}t \, \left[v_x(t)\frac{\delta B_y(t)}{B_0} - v_y(t)\frac{\delta B_x(t)}{B_0}\right]. \end{aligned} \tag{6.5}$$

Within QLT, we replace the particle properties on the right-hand side of (6.5) by unperturbed orbits. In this case, we find for the first-order velocity correlation function

$$\begin{aligned}
V_{xj}^{(1)}(\tau) &= v_x(0)v_j(0) + \Omega v_j(0)\int_0^\tau \mathrm{d}t\; v_y(t), \\
&= v_x(0)v_j(0) + \Omega v_j(0) v\sqrt{1-\mu^2}\int_0^\tau \mathrm{d}t\; \sin(\Phi_0 - \Omega t), \\
&= v\sqrt{1-\mu^2} v_j(0)\cos(\Phi_0 - \Omega t), \\
&= v_x(t)v_j(0),
\end{aligned} \tag{6.6}$$

because $\langle \delta B_i \rangle = 0$. In general, it can be demonstrated that

$$V_{ij}^{(1)}(t) = \big\langle v_i(t)v_j(0)\big\rangle = \big[v_i(t)v_j(0)\big]_{\mathrm{QLT}} = V_{ij}^{(0)}(t). \tag{6.7}$$

With (3.32), it can easily be shown that

$$\sum_{i=x,y} V_{ii}^{(1)}(t) = v^2(1-\mu^2)\,\cos(\Omega t). \tag{6.8}$$

This result can be used to substitute the velocity correlation function in (6.4).

6.1.2 The Time-dependent Pitch-angle Fokker–Planck Coefficient

Before we calculate the characteristic function, it is useful to determine the time-dependent quasilinear Fokker–Planck coefficient $D_{\mu\mu}^{(1)}(t)$, which can be defined as

$$D_{\mu\mu}^{(1)}(t) = \frac{\big\langle (\Delta\mu(t))^2 \big\rangle}{2t}. \tag{6.9}$$

The time-independent Fokker–Planck coefficient can then simply be calculated from

$$D_{\mu\mu}^{(1)} = \lim_{t\to\infty} D_{\mu\mu}^{(1)}(t). \tag{6.10}$$

To proceed, we apply again the equation of motion (e.g., (3.3)) to obtain

$$\Delta\mu(\xi) = \mu(\xi) - \mu(0) = \frac{\Omega}{vB_0}\int_0^\xi \mathrm{d}t\; \big(v_x\delta B_y - v_y\delta B_x\big)(t). \tag{6.11}$$

Therefore, the time-dependent Fokker–Planck coefficient can be written as

$$D_{\mu\mu}^{(1)}(\xi) = \frac{1}{2\xi}\frac{\Omega^2}{v^2 B_0^2} \times \int_0^\xi \mathrm{d}t_1 \int_0^\xi \mathrm{d}t_2 \left\langle \left(v_x \delta B_y - v_y \delta B_x\right)(t_1) \left(v_x \delta B_y^* - v_y \delta B_x^*\right)(t_2)\right\rangle. \tag{6.12}$$

To proceed, we apply QLT to replace the velocity components v_i (see (1.10))

$$D_{\mu\mu}^{(1)}(\xi) = \frac{1}{2\xi}\frac{\Omega^2(1-\mu^2)}{B_0^2}\int_0^\xi \mathrm{d}t_1 \int_0^\xi \mathrm{d}t_2 \left[\cos(\Omega t_1)\cos(\Omega t_2) + \sin(\Omega t_1)\sin(\Omega t_2)\right] R_{xx}(t_1,t_2), \tag{6.13}$$

with $R_{xx}(t_1,t_2) = \langle \delta B_x(t_1)\delta B_x^*(t_2)\rangle$. By using the common relation

$$\cos(x)\cos(y) + \sin(x)\sin(y) = \cos(x-y), \tag{6.14}$$

it follows

$$D_{\mu\mu}^{(1)}(\xi) = \frac{1}{2\xi}\frac{\Omega^2(1-\mu^2)}{B_0^2}\int_0^\xi \mathrm{d}t_1 \int_0^\xi \mathrm{d}t_2 \cos\left[\Omega(t_1-t_2)\right] R_{xx}(t_1,t_2). \tag{6.15}$$

Here, we used again $\Phi_0 = 0$ for the initial gyrophase due to the assumption of axisymmetric turbulence. For the slab model, we have

$$R_{xx}(t_1,t_2) = 4\pi \int_0^\infty \mathrm{d}k_\parallel \, g^{\mathrm{slab}}(k_\parallel)\, \mathrm{e}^{\mathrm{i}k_\parallel v\mu(t_1-t_2)} \tag{6.16}$$

and, therefore,

$$D_{\mu\mu}^{(1)}(\xi) = \frac{2\pi\Omega^2(1-\mu^2)}{\xi B_0^2}\int_0^\infty \mathrm{d}k_\parallel \, g^{\mathrm{slab}}(k_\parallel) \times \int_0^\xi \mathrm{d}t_1 \int_0^\xi \mathrm{d}t_2 \cos\left[\Omega(t_1-t_2)\right] \mathrm{e}^{\mathrm{i}k_\parallel v\mu(t_1-t_2)}. \tag{6.17}$$

The two time-integrals are elementary and we find

$$D_{\mu\mu}^{(1)}(t) = \frac{2\pi\Omega^2(1-\mu^2)}{B_0^2}\int_0^\infty \mathrm{d}k_\parallel \, g^{\mathrm{slab}}(k_\parallel)\left[R_+^{(1)}(t) + R_-^{(1)}(t)\right], \tag{6.18}$$

with the resonance function

$$R_{\pm}^{(1)}(t) = \frac{1-\cos(\beta_{\pm}t)}{t\beta_{\pm}^2} \tag{6.19}$$

and the resonance condition $\beta_{\pm} = k_{\parallel} v\mu \pm \Omega$.

6.1.3 The Ensemble Averaged Parallel Position

The quasilinear ensemble averaged parallel position $\langle z \rangle^{(1)}$ can also be obtained from the equations of motion:

$$\langle z(\xi) \rangle^{(1)} = v\mu\xi + \frac{\Omega}{B_0}\int_0^{\xi} \mathrm{d}\tau \int_0^{\tau} \mathrm{d}t \, \left\langle (v_x \delta B_y - v_y \delta B_x)(t) \right\rangle . \tag{6.20}$$

Within QLT, v_x and v_y are unperturbed quantities. With $\langle \delta B_x \rangle = \langle \delta B_y \rangle = 0$, we therefore find

$$\langle z(\xi) \rangle^{(1)} = v\mu\xi. \tag{6.21}$$

This is the expected QLT result for the ensemble averaged parallel position. Within QLT, the ensemble average moves unperturbed through the turbulence.

6.1.4 The Quasilinear Mean Square Displacement

Here, we calculate the quasilinear time-dependent mean square displacement $\sigma_z^2 = \langle z^2(\xi) \rangle - \langle z(\xi) \rangle^2$. In the following, we neglect the index in the mean square deviation ($\sigma_z \equiv \sigma_z^{(1)}$). From the equations of motion, we obtain

$$\begin{aligned} \langle z^2(\xi) \rangle = (v\mu\xi)^2 + \frac{\Omega^2}{B_0^2} \int_0^{\xi} \mathrm{d}\tau_1 \int_0^{\xi} \mathrm{d}\tau_2 \\ \times \int_0^{\tau_1} \mathrm{d}t_1 \int_0^{\tau_2} \mathrm{d}t_2 \left\langle (v_x \delta B_y - v_y \delta B_x)(t_1) \cdot (v_x \delta B_y^* - v_y \delta B_x^*)(t_2) \right\rangle \end{aligned} \tag{6.22}$$

and, therefore,

$$\begin{aligned} \sigma_z^2(\xi) = \frac{\Omega^2}{B_0^2} \int_0^{\xi} \mathrm{d}\tau_1 \int_0^{\xi} \mathrm{d}\tau_2 \int_0^{\tau_1} \mathrm{d}t_1 \int_0^{\tau_2} \mathrm{d}t_2 \\ \times \left\langle (v_x \delta B_y - v_y \delta B_x)(t_1)(v_x \delta B_y^* - v_y \delta B_x^*)(t_2) \right\rangle . \end{aligned} \tag{6.23}$$

By applying QLT and by repeating the steps made in the previous paragraph, we find

$$\sigma_z^2(\xi) = \frac{\Omega^2 v^2 (1-\mu^2)}{B_0^2} \int_0^\xi d\tau_1 \int_0^\xi d\tau_2 \int_0^{\tau_1} dt_1 \int_0^{\tau_2} dt_2 \times \left[\cos(\Omega t_1)\cos(\Omega t_2) + \sin(\Omega t_1)\sin(\Omega t_2)\right] R_{xx}(t_1, t_2), \tag{6.24}$$

where we used again $\Phi_0 = 0$ and $R_{xx}(t_1, t_2) = \langle \delta B_x(t_1) \delta B_x^*(t_2) \rangle$. For $R_{xx}(t_1, t_2)$, we can use (6.16) to find

$$\sigma_z^2(\xi) = \frac{4\pi \Omega^2 v^2 (1-\mu^2)}{B_0^2} \int_0^\infty dk_\parallel \, g^{\text{slab}}(k_\parallel) \times \int_0^\xi d\tau_1 \int_0^\xi d\tau_2 \int_0^{\tau_1} dt_1 \int_0^{\tau_2} dt_2 \, \cos\left[\Omega(t_1 - t_2)\right] e^{ik_\parallel v\mu(t_1 - t_2)}. \tag{6.25}$$

These integrals are elementary, and we obtain

$$\sigma_z^2(t) = \frac{4\pi v^2 \Omega^2 (1-\mu^2)}{B_0^2} \int_0^\infty dk_\parallel \, g^{\text{slab}}(k_\parallel) \left[M_+^{(1)}(t) + M_-^{(1)}(t)\right]. \tag{6.26}$$

In (6.26) we used the resonance function

$$M_\pm^{(1)}(t) = \frac{1 - \cos(\beta_\pm t)}{\beta_\pm^4} - \frac{\sin(\beta_\pm t)}{\beta_\pm^3} t + \frac{t^2}{2\beta_\pm^2} \tag{6.27}$$

and the resonance condition $\beta_\pm = k_\parallel v\mu \pm \Omega$. The time derivative of this function is

$$\frac{dM_\pm^{(1)}}{dt} = \frac{1 - \cos(\beta_\pm t)}{\beta_\pm^2} t = t^2 R_\pm^{(1)}(t), \tag{6.28}$$

with $R_\pm^{(1)}(t)$ from (6.19). Therefore, we can derive the useful relation

$$\frac{d}{dt}\sigma_z^2 = 2v^2 t^2 D_{\mu\mu}^{(1)}(t), \tag{6.29}$$

where the first-order Fokker–Planck coefficient is given by (6.18), allowing this result to be rewritten as an integral representation

$$\sigma_z^2(t) = 2v^2 \int_0^t d\tau \left[\tau^2 D_{\mu\mu}^{(1)}(\tau)\right], \tag{6.30}$$

where $\sigma_z^2(0) = 0$ has been used.

6.2 The Resonance Function of SOQLT

For the characteristic function, we have within SOQLT ((6.21) combined with (6.2))

$$\Gamma^{(1)}(\mathbf{k}) = \mathrm{e}^{\mathrm{i}k_\parallel \langle z\rangle} \mathrm{e}^{-\frac{\sigma_z^2(t)}{2}k_\parallel^2} = \mathrm{e}^{\mathrm{i}k_\parallel v\mu t} \mathrm{e}^{-\frac{\sigma_z^2(t)}{2}k_\parallel^2}. \tag{6.31}$$

Together with (6.1), we then obtain

$$D_{\mu\mu}^{(2)} = \frac{2\pi\Omega^2(1-\mu^2)}{B_0^2} \int_0^\infty \mathrm{d}k_\parallel\, g^{\mathrm{slab}}(k_\parallel) \left[R_{+1}^{(2)}(k_\parallel) + R_{-1}^{(2)}(k_\parallel)\right], \tag{6.32}$$

where we introduced the resonance function of SOQLT:

$$R_n^{(2)} := \int_0^\infty \mathrm{d}t\, \cos\left[(k_\parallel v\mu + n\Omega)t\right] \mathrm{e}^{-\frac{\sigma_z^2(t)}{2}k_\parallel^2}. \tag{6.33}$$

For $\sigma_z^2(t)$, we can use (6.26) or (6.30). In the following, we consider two mathematical approximations to simplify this form of the resonance function.

6.2.1 *The 90°-Approximation*

Because we are mainly interested in pitch-angles close to 90°, we approximate the parameter σ_z^2 by

$$\sigma_z^2(t,\mu) \approx \sigma_z^2(t,\mu=0). \tag{6.34}$$

For $\mu = 0$, we can deduce from (6.27)

$$M_+(t) + M_-(t) = \frac{m(\Omega t)}{\Omega^4}, \tag{6.35}$$

with

$$m(\Omega t) = 2 - 2\cos(\Omega t) - 2\sin(\Omega t)\cdot(\Omega t) + (\Omega t)^2. \tag{6.36}$$

Hence, (6.26) becomes

$$\sigma_z^2(t,\mu=0) = \frac{4\pi v^2}{\Omega^2 B_0^2} m(\Omega t) \int_0^\infty \mathrm{d}k_\parallel\, g^{\mathrm{slab}}(k_\parallel). \tag{6.37}$$

By making use of

$$\int_0^\infty dk_\parallel g^{\text{slab}}(k_\parallel) = \frac{1}{8\pi}\delta B_{\text{slab}}^2, \tag{6.38}$$

we obtain

$$\sigma_z^2(t, \mu = 0) = \frac{v^2}{2}\frac{\delta B^2}{B_0^2}\frac{m(\Omega t)}{\Omega^2} = \frac{R_L^2}{2}\frac{\delta B^2}{B_0^2}m(\Omega t). \tag{6.39}$$

This is the exact QLT result for $\mu = 0$. Within the 90°-approximation, we use

$$\sigma_z^2(t, \mu) \approx \frac{R_L^2}{2}\frac{\delta B^2}{B_0^2}m(\Omega t). \tag{6.40}$$

The justification for this approximation is as follows: QLT is assumed to be correct for $\mu \neq 0$. In this case, however, the corrections described by (6.40) are quite small due to the dominant gyroresonant interaction and can, therefore, be neglected. Thus, the failure of (6.40) at $\mu \neq 0$ is negligible. For $\mu = 0$, however, the corrections to the unperturbed orbit are essential. Thus, we need an approximation for $\sigma_z^2(t, \mu)$ that is accurate at 90°. These requirements are fulfilled by the 90°-approximation.

6.2.2 *The 90°-Late-time Approximation*

To achieve a further simplification of the mean square deviation and the resonance function, we approximate σ_z^2 by its value at 90° and for times larger than the inverse gyrofrequency:

$$\sigma_z^2(t, \mu) \approx \sigma_z^2(\Omega t \gg 1, \mu = 0) = \frac{v^2}{2}\frac{\delta B^2}{B_0^2}t^2. \tag{6.41}$$

The LTA (large-time or late-time approximation) is correct only if small time scales ($t \ll \Omega^{-1}$) are unimportant. This assumption is comparable to the diffusion approximation that has been applied during the derivation of the NLGC theory (see Chap. 4) and WNLT (see (5.15)).

6.3 Comparison with Previous Theories

Several nonlinear theories have been proposed in the past (see e.g., Chap. 4) to describe pitch-angle diffusion. Now we want to compare SOQLT with these previous theories to find out whether their results can be recovered by SOQLT.

6.3.1 The Nonlinear Perturbation Theory

One of the first attempts to solve the 90°-problem had been proposed by Völk (1973) by using a nonlinear perturbation theory (NPT) based on the methods originally developed by Dupree (1966, 1967). Within the NPT of Völk (1973), the resonance function has the form

$$R_n^{\mathrm{NPT}} = \int_0^\infty \mathrm{d}t \, \cos\left[(k_\parallel v\mu + n\Omega)t\right] \, \mathrm{e}^{-\frac{v^2}{3}D_{\mu\mu}k_\parallel^2 t^3}. \tag{6.42}$$

Equation (6.42) can also be obtained from second-order QLT. If we replace the time-dependent first-order Fokker–Planck coefficient in (6.30) by the time-independent coefficient

$$D_{\mu\mu}^{(1)}(\tau) \approx D_{\mu\mu}^{(1)}, \tag{6.43}$$

we find for the mean square deviation

$$\sigma_z^2(t) = 2v^2 D_{\mu\mu}^{(1)} \int_0^t \mathrm{d}\tau \, \tau^2 = \frac{2v^2}{3} t^3 D_{\mu\mu}^{(1)}. \tag{6.44}$$

Together with (6.33), we find a result that is exactly equal to Völk's nonlinear perturbation theory (6.42). In this book, the approximation of (6.43) is referred to as the *diffusion approximation.*

However, by exploring the NPT numerically and analytically, one still obtains $\tilde{D}_{\mu\mu}(\mu = 0) = 0$. This nonvanishing result obtained by Völk (1973) comes due to the crude approximations used therein. Völk derived a formula for $D_{\mu\mu}$ that yields $\tilde{D}_{\mu\mu}(\mu = 0) \approx 1.1 \times 10^{-5}$ for the parameter set considered in the simulations (see (3.79)), which is clearly in disagreement with the simulations ($\tilde{D}_{\mu\mu}(\mu = 0) \approx 0.6 \times 10^{-3}$). Obviously, the nonlinear perturbation theory based on the work of Dupree (1966) cannot reproduce neither the test-particle simulations nor can the theory solve the 90°-problem. Even a qualitative solution of the 90°-problem (nonvanishing pitch angle scattering) cannot be achieved by this approach.

6.3.2 The Partially Averaged Field Theory

Here, we compare SOQLT with the partially averaged field theory (PAFT) of Jones et al. (1973, 1978). Within the 90°-LTA, we have

$$\langle z^2 \rangle = (v\mu t)^2 + \frac{v^2}{2} \frac{\delta B^2}{B_0^2} t^2. \tag{6.45}$$

As a (wrong) approximation, one could use

$$\langle z\rangle \approx \sqrt{\langle z^2\rangle} = vt\sqrt{\mu^2 + \frac{1}{2}\frac{\delta B^2}{B_0^2}}, \tag{6.46}$$

which is exactly the result of the PAFT of Jones et al. (1973, 1978) with $\alpha_{\mathrm{NL}} = (1-\mu^2)/2$. Obviously, the approximation of (6.46) is not correct. The additional effects described by the SOQLT result in the broadening of the Gaussian function ($\sigma_z^2 \neq 0$) and not in a different average position.

6.3.3 *The Heuristic Ansatz by Völk*

After realizing that the NPT of Völk (1973) cannot solve the 90°-problem, Völk (1975) included the effect of particle trapping directly into transport theory. By using some approximations, Völk used the following *ansatz* for the resonance function:

$$R_n \approx \begin{cases} \frac{\sqrt{\pi}B_0}{v\delta B k_\parallel} & \text{for} \quad | k_\parallel v\mu + n\Omega | \ll \frac{v\delta B k_\parallel}{B_0}, \\ 0 & \text{otherwise.} \end{cases}$$

Although heuristic, this form for the resonance function can be applied to calculate pitch-angle diffusion coefficients. If we apply the 90°-LTA, the resonance function of SOQLT becomes ((6.41) with (6.33))

$$\begin{aligned} K_n^{(2)} &= \int_0^\infty dt\ \cos\left[(k_\parallel v\mu + n\Omega)t\right] e^{-\frac{v^2}{4}\frac{\delta B^2}{B_0^2}t^2k_\parallel^2}, \\ &= \frac{\sqrt{\pi}B_0}{v\delta B k_\parallel} \exp\left[-\left(\frac{(k_\parallel v\mu + n\Omega)B_0}{v\delta B k_\parallel}\right)^2\right]. \end{aligned} \tag{6.47}$$

If one applies the approximation originally proposed by Dupree (1966), he finds

$$R_n^{(2)} \approx \begin{cases} \frac{\sqrt{\pi}B_0}{v\delta B k_\parallel} & \text{for} \quad | k_\parallel v\mu + n\Omega | \ll \frac{v\delta B k_\parallel}{B_0}, \\ 0 & \text{otherwise,} \end{cases} \tag{6.48}$$

which agrees with the ansatz of Völk (1975, (48)). It is an interesting matter that we obtain the theories of Völk (1973, 1975) and Jones et al. (1973, 1978) as limiting cases of SOQLT calculations.

6.3.4 *The Strong Turbulence, Weak Coupling Theory*

Based on the work by Kubo (1962), Weinstock (1969, 1970), and Ben-Israel et al. (1975), Goldstein (1976) developed and applied a strong turbulence, weak coupling theory to explore the 90°-problem. In contrast to the theories discussed in the previous paragraphs, the approach of Goldstein does not allow for the derivation of simple formulas even in the case of magnetostatic slab turbulence and a simplified form of the wave spectrum. Therefore, a rigorous discussion of the strong turbulence, weak coupling theory is not presented here. In the following, only the most important ideas of the theory and the results are summarized. Goldstein (1976) stated that, in his theory, the computed particle trajectories are more general, in comparison to the theories proposed by Völk (1973, 1975), Jones et al. (1973), and Owens (1974). According to Goldstein (1976), the strong turbulence, weak coupling theory agrees with the simulations of Kaiser (1975). The main disadvantage of the strong turbulence, weak coupling theory is that, for realistic wave spectra and turbulence models (e.g., dynamical turbulence models), the theory is analytically and numerically quite intractable.

6.4 Analytical Results of SOQLT

Here, we apply the results of the previous sections to calculate the second-order Fokker–Planck coefficient $D^{(2)}_{\mu\mu}$. For convenience, we define the dimensionless parameters

$$
\begin{aligned}
R &= \frac{R_{\mathrm{L}}}{l_{\mathrm{slab}}} = \frac{v}{\Omega l_{\mathrm{slab}}}, \\
\rho &= k_{\mathrm{slab}} l_{\mathrm{slab}}, \\
\tilde{D}_{\mu\mu} &= \frac{l_{\mathrm{slab}}}{v} D_{\mu\mu}, \\
\tilde{\sigma}_z^2 &= \frac{\sigma_z^2}{l_{\mathrm{slab}}^2}.
\end{aligned}
\tag{6.49}
$$

With these parameters and by applying the integral transformations $x = k_{\parallel} l_{\mathrm{slab}}$ and $\tau = \Omega t$, we obtain for the resonance function

$$
R_n^{(2)}(x) = \frac{1}{\Omega} \int_0^{\infty} \mathrm{d}\tau \, \cos\left[(\mu R x + n)\tau\right] \, \mathrm{e}^{-\frac{\tilde{\sigma}_z^2(\tau)}{2} x^2}. \tag{6.50}
$$

With the parameters from (6.49), the dimensionless second-order pitch-angle Fokker–Planck coefficient becomes

$$\tilde{D}^{(2)}_{\mu\mu} = \frac{C(\nu)(1-\mu^2)\Omega}{R}\frac{\delta B^2}{B_0^2}\int_0^\infty \mathrm{d}x\, h(x,\rho_{\mathrm{slab}},\nu,p)\left[R_+^{(2)}(x)+R_-^{(2)}(x)\right], \tag{6.51}$$

where the dimensionless wave spectrum $h(x)$ defined as

$$g^{\mathrm{slab}}(k_\parallel) = \frac{C(\nu)}{2\pi} l_{\mathrm{slab}}\delta B^2\, h(x = k_\parallel l_{\mathrm{slab}}) \tag{6.52}$$

has been used. With the resonance function from (6.50), it follows that

$$\tilde{D}^{(2)}_{\mu\mu} = \frac{C(\nu)(1-\mu^2)}{R}\frac{\delta B^2}{B_0^2}\int_0^\infty \mathrm{d}x\, h(x,\rho_{\mathrm{slab}},\nu,p) \times \int_0^\infty \mathrm{d}\tau\, \mathrm{e}^{-\frac{\tilde{\sigma}_z^2(\tau)}{2}x^2}\left\{\cos\left[(\mu R x+1)\tau\right]+\cos\left[(\mu R x-1)\tau\right]\right\}. \tag{6.53}$$

This Fokker–Planck coefficient can be calculated numerically, which will be done in Sect. 6.5. In the next paragraph, several forms of the wave spectrum are discussed.

6.4.1 Different Forms of the Wave Spectrum

In the following, we calculate transport parameters for different forms of the wave spectrum $g^{\mathrm{slab}}(k_\parallel)$:

- The standard spectrum used so far in this book (2.33). Here we have

$$h(x,\nu) = (1+x^2)^{-\nu}. \tag{6.54}$$

- The cutoff-spectrum, which has the form

$$g^{\mathrm{slab}}(k_\parallel) = \frac{C(\nu)}{2\pi} l_{\mathrm{slab}}\delta B^2 \begin{cases} (1+k_\parallel^2 l_{\mathrm{slab}}^2)^{-\nu} & \text{for } k_\parallel \le k_{\mathrm{d,slab}}, \\ 0 & \text{for } k_\parallel \ge k_{\mathrm{d,slab}}, \end{cases} \tag{6.55}$$

  with $l_{\mathrm{slab}} k_{\mathrm{d,slab}} \gg 1$. Hence, we have

$$h(x,\rho,\nu) = \begin{cases} (1+x^2)^{-\nu} & \text{for } x \le \rho, \\ 0 & \text{for } x \ge \rho. \end{cases} \tag{6.56}$$

- A spectrum with dissipation range is given by

$$g^{\text{slab}}(k_\parallel) = \frac{C(\nu)}{2\pi} l_{\text{slab}} \delta B^2 \times \begin{cases} (1 + k_\parallel^2 l_{\text{slab}}^2)^{-\nu} & \text{for } k_\parallel \leq k_{\text{d,slab}}, \\ (1 + k_{\text{d,slab}}^2 l_{\text{slab}}^2)^{-\nu} (k_{\text{d,slab}}/k_\parallel)^p & \text{for } k_\parallel \geq k_{\text{d,slab}}. \end{cases} \tag{6.57}$$

By comparing this spectrum with the heliospheric measurements shown in Fig. 2.3, we find that this spectrum provides the best fit for the observations. In addition to the parameters used in the standard spectrum, the dissipation wavenumber $k_{d,\text{slab}}$ and the dissipation range spectral index p have been used. Here we have,

$$h(x, \rho, \nu, p) = \begin{cases} (1 + x^2)^{-\nu} & \text{for } x \leq \rho, \\ (1 + \rho^2)^{-\nu} \left(\frac{\rho}{x}\right)^p & \text{for } x \geq \rho. \end{cases} \tag{6.58}$$

It can easily be seen by choosing $k_\text{d} \to \infty$ that we can obtain the standard spectrum from (6.57). The cutoff spectrum can be obtained from the Bieber et al. (1994) spectrum in the limit $p \to \infty$. It should be emphasized that magnetostatic QLT yields $\lambda_\parallel = \infty$ for the cutoff spectrum as well as for the improved spectrum of (6.57).

6.4.2 *Analytical Results for 90°-Scattering*

To obtain analytical results, we must consider the 90°-late-time approximation. Only within this approximation, analytical results for the Fokker–Planck coefficients can be obtained. Combining (6.41) with (6.53), we have

$$\tilde{D}_{\mu\mu}^{(2)}(\mu) = \frac{C(\nu)(1-\mu^2)}{R} \frac{\delta B^2}{B_0^2} \int_0^\infty \text{d}x \, h(x) \times \int_0^\infty \text{d}\tau \, \text{e}^{-\frac{\gamma^2}{4} x^2 \tau^2} \{\cos[(\mu R x + 1)\tau] + \cos[(\mu R x - 1)\tau]\}, \tag{6.59}$$

with $\gamma = R\delta B / B_0$. By using

$$\int_0^\infty \text{d}\tau \cos[(\mu R x \pm 1)\tau] \, \text{e}^{-\frac{\gamma^2}{4} x^2 \tau^2} = \frac{\sqrt{\pi}}{\gamma x} \text{e}^{-\frac{(\mu R x \pm 1)^2}{\gamma^2 x^2}}, \tag{6.60}$$

we find

$$\tilde{D}_{\mu\mu}^{(2)}(\mu) = \frac{\sqrt{\pi} C(\nu)(1-\mu^2)\delta B}{R^2 B_0} \int_0^\infty \frac{\text{d}x}{x} h(x) \left[\text{e}^{-\frac{(\mu R x + 1)^2}{\gamma^2 x^2}} + \text{e}^{-\frac{(\mu R x - 1)^2}{\gamma^2 x^2}}\right]. \tag{6.61}$$

To proceed, it is necessary to specify the wave spectrum $h(x)$ and to apply further approximations. In turn, we consider analytical results of (6.61) for $\mu = 0$, using the standard spectrum of (6.54). Then we obtain

$$\tilde{D}^{(2)}_{\mu\mu}(\mu = 0) = \frac{2\sqrt{\pi}C(\nu)\delta B}{R^2 B_0} \int_0^\infty \frac{\mathrm{d}x}{x}(1 + x^2)^{-\nu}\,\mathrm{e}^{-(\gamma x)^{-2}}. \tag{6.62}$$

Applying the integral transformation $y = x^{-2}$ and splitting the integral, we find approximately

$$\tilde{D}^{(2)}_{\mu\mu}(\mu = 0) \approx \frac{\sqrt{\pi}C(\nu)\delta B}{R^2 B_0} \left\{ \int_0^1 \mathrm{d}y\; y^{\nu-1}\,\mathrm{e}^{-y/\gamma^2} + \int_1^\infty \mathrm{d}y\; y^{-1}\,\mathrm{e}^{-y/\gamma^2} \right\}. \tag{6.63}$$

These two integrals can be solved, and we obtain

$$\tilde{D}^{(2)}_{\mu\mu}(\mu = 0) \approx \frac{\sqrt{\pi}C(\nu)\delta B}{R^2 B_0} \left[\gamma^{2\nu}\Gamma(\nu) - \gamma^{2\nu}\Gamma(\nu, \gamma^{-2}) + E_1(\gamma^{-2})\right], \tag{6.64}$$

where we used the incomplete gamma function $\Gamma(x, y)$ and the exponential integral function $E_1(x)$ (see Abramowitz and Stegun 1974). For the parameter regime considered in numerical simulations, γ^{-1} is a large number. If we apply (Abramowitz and Stegun 1974)

$$\begin{aligned} \mathrm{E}_1(\gamma^{-2} \gg 1) &= \Gamma(0, \gamma^{-2} \gg 1) \approx \gamma^2\,\mathrm{e}^{-\gamma^{-2}} \\ \Gamma(\nu, \gamma^{-2} \gg 1) &\approx \gamma^{2-2\nu}\,\mathrm{e}^{-\gamma^{-2}}, \end{aligned} \tag{6.65}$$

we find

$$\tilde{D}^{(2)}_{\mu\mu}(\mu = 0) \approx \sqrt{\pi}\Gamma(\nu)C(\nu)R^{2\nu-2}\left(\frac{\delta B}{B_0}\right)^{2\nu+1}. \tag{6.66}$$

This formula describes second-order pitch-angle diffusion at 90° for $R\delta B/B_0 \ll 1$. Obviously, we no longer find $D_{\mu\mu} \sim (\delta B/B_0)^2$. Furthermore, the rigidity dependence is the same as in standard quasilinear theory. Thus, we expect that for the standard spectrum used here, the energy dependence of the parallel mean free paths is the same as in QLT. This statement is in accord with Fig. 3.2.

6.5 Numerical Results for Fokker–Planck Coefficients and Mean Free Paths

Here, the results of the previous sections are applied to calculate numerically the second-order pitch-angle Fokker–Planck coefficient and the second-order parallel mean free path.

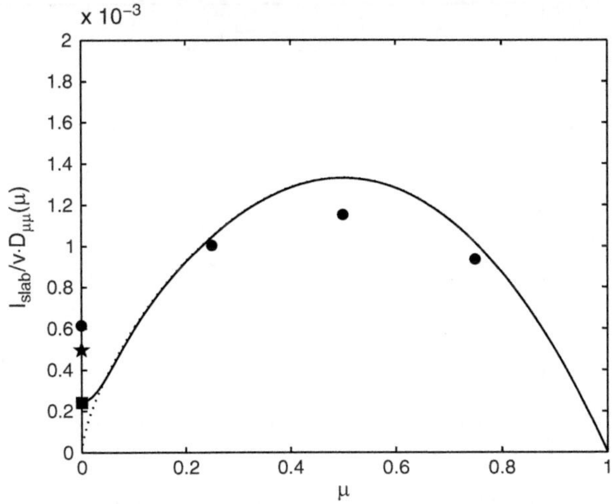

Fig. 6.1 Shown are the results of QLT (*dotted line*), SOQLT within the 90°-LTA (*solid line*), and simulations (Qin 2002, dots) for $D_{\mu\mu}$. Also shown is the more accurate SOQLT result without LTA (*star*), which is closer to the simulations. Furthermore, the analytical result from (6.66) is shown (*square*). In this plot, the rigidity is $R = 0.0363$ and the strength of turbulence is assumed to be $\delta B/B_0 = 0.05$

6.5.1 Numerical Results for $D_{\mu\mu}^{(2)}$

Figure 6.1 shows the pitch-angle Fokker–Planck coefficient for $R = 0.0363$, $2\nu = 5/3$, and $\delta B/B_0 = 0.05$ for the standard spectrum. The well-known QLT result is compared with the simulations of Qin (2002) and SOQLT results. The SOQLT results are calculated by using the 90°-approximation and the 90°-LTA (late-time approximation) by integrating (6.53) numerically. Because of the difference between these two approximations, one comes to the conclusion that early times are important if we like to achieve agreement with simulations for $\mu = 0$. In Fig. 6.2, it is shown that for stronger turbulence ($\delta B/B_0 = 0.25$), the nonlinear effects become more and more important. In Fig. 6.3, the Fokker–Planck coefficients are compared for $\delta B/B_0 = 1$, which should be a realistic value in the solar wind at 1 AU heliocentric distance. For such moderate turbulence, we find a nearly isotropic form of the pitch-angle Fokker–Planck coefficient $D_{\mu\mu} \approx D \cdot (1 - \mu^2)$ and, thus, the pitch-angle Fokker–Planck coefficient has its maximum at 90°.

6.5.2 Numerical Results for $\lambda_{\parallel}^{(2)}$

Figure 6.4 shows the parallel mean free path calculated with QLT and SOQLT within the 90°-LTA. To obtain the numerical results for $\lambda_{\parallel}$, (6.61) is combined with (3.1).

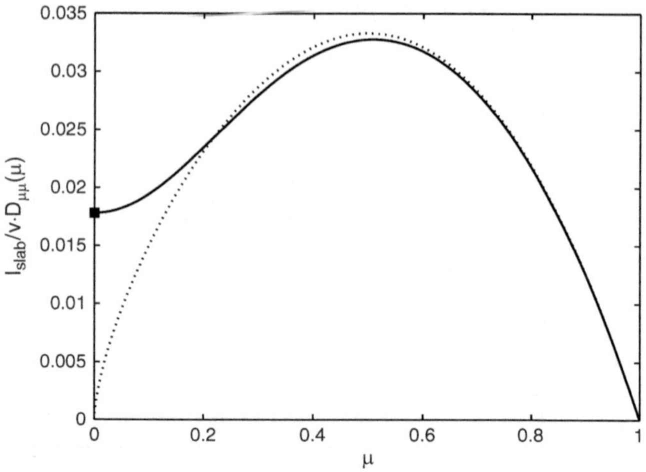

Fig. 6.2 The pitch-angle Fokker–Planck coefficient $D_{\mu\mu}$, calculated with QLT (*dotted line*) and SOQLT with LTA (*solid line*). Also the analytical result from (6.66) is shown (*square*). Here, the rigidity is $R = 0.0363$ and the strength turbulence is $\delta B / B_0 = 0.25$

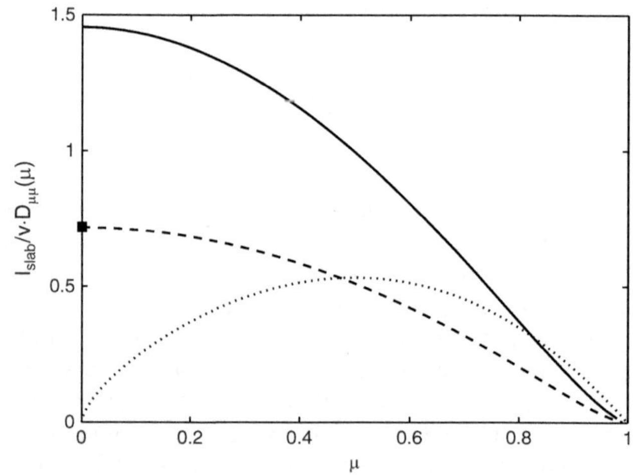

Fig. 6.3 The pitch-angle Fokker–Planck coefficient $D_{\mu\mu}$ calculated with QLT (*dotted line*), SOQLT with (*dashed line*) and without (*solid line*) LTA. Also the analytical result from (6.66) is shown (*square*). The rigidity is $R = 0.0363$ and the strength of turbulence is $\delta B / B_0 = 1$. The second-order results are nearly isotropic in the pitch-angle ($D_{\mu\mu} \sim (1 - \mu^2)$)

The integrals over x and μ herein are solved numerically. SOQLT is applied to calculate the parallel mean free path for the standard spectrum ($k_{\text{slab}} \to \infty$), the cutoff-spectrum ($p \to \infty$), and the spectrum with dissipation range and with $p = 3$. For the slab bendover scale, we assume $l_{\text{slab}} = 0.03\,\text{AU}$, and for the dissipation wavenumber, we assume $k_{\text{slab}} = 3 \times 10^6\,\text{AU}^{-1}$.

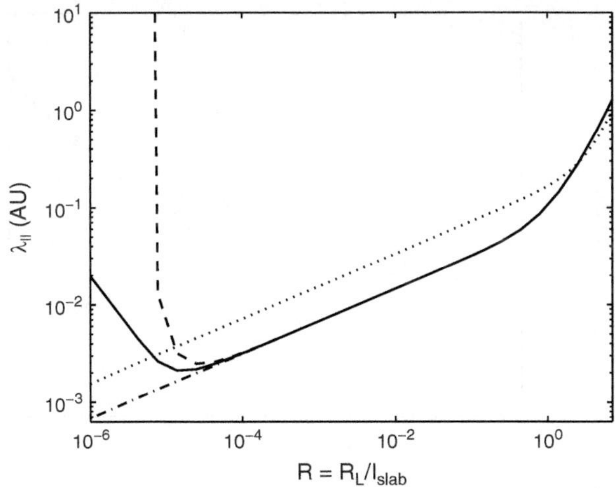

Fig. 6.4 The parallel mean free path as a function of the rigidity R for $\delta B/B_0 = 1$. Shown are dissipationless QLT results (*dotted line*) in comparison with SOQLT results for a dissipationless spectrum (*dash-dotted line*), a spectrum with cutoff (*dashed line*), and a spectrum with dissipation range spectral index $p = 3$ (*solid line*)

As expected, the parallel mean free path is no longer infinitely large within SOQLT for the two spectra with cutoff and dissipation effects. For the standard spectrum, the deviation between QLT and SOQLT is small. But clearly the nonlinear effect can be seen, resulting in a smaller parallel mean free path. For the cutoff spectrum and for the spectrum with dissipation range, we find a quite large parallel mean free path at smaller rigidities. The QLT result for these two spectra is $\lambda_{\|} = \infty$. Because of the strong nonlinear effect, one would expect that QLT is not appropriate to reproduce observed parallel mean free paths in the solar wind. However, comparable strong 90°-scattering effects can be found for dynamical turbulence. For such improved turbulence models, it may be possible that QLT is recovered due to the supression of nonlinear effects. This possibility will be further discussed in Chap. 8, where it is demonstrated that a combination of QLT and dynamical turbulence models can reproduce heliospheric observations of $\lambda_{\|}$.

It is interesting to explore the validity of SOQLT for moderate ($\delta B^2/B_0^2 = 1$) slab turbulence to see the strength of the 90°-scattering effect. For the second-order results, the 90°-LTA is applied to simplify the calculations of the parallel mean free path. As demonstrated in Fig. 6.5, we can see the nonlinear effect if we assume moderate strong turbulence. The agreement between SOQLT and the simulations is excellent, but even the quasilinear result does not deviate too much: $\lambda_{\|}^{\rm QLT}/\lambda_{\|}^{\rm Sim}(R = 0.04) \approx 2$.

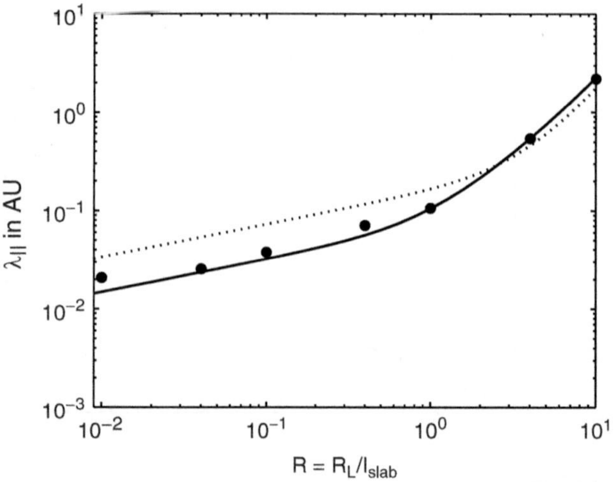

Fig. 6.5 $\lambda_{\parallel}$ for pure slab geometry and moderate turbulence ($\delta B_{\mathrm{slab}}^2/B_0^2 = 1$). The simulations are compared with QLT and SOQLT. Obviously SOQLT agrees much better with the simulations than QLT

6.5.3 Steep Wave Spectra

In Lazar et al. (2003) and Spanier and Schlickeiser (2005), the heating rate of the interstellar medium (ISM), especially the warm ionized medium, has been calculated. In these two articles, it was demonstrated that a steeper form of the wave spectrum ($s = 2\nu > 2$) could be reasonable. Although such a steep behavior of the wave spectrum cannot be confirmed by observations (see Chap. 2), it is an interesting matter to explore parallel diffusion of cosmic rays for such steep wave spectra. To explore steep wave spectra and to test SOQLT, we consider simulations for slab geometry. Two different values for the inertial range spectral index are investigated, namely $s = 2.5$ and $s = 2 + 5/3$. Because of the 90°-problem, magnetostatic QLT cannot be applied for such turbulence parameters (the QLT result is $\lambda_{\parallel}^{\mathrm{QLT}} = \infty$). As demonstrated in Fig. 6.6, the second-order parallel mean free path decreases with increasing rigidity. Obviously, SOQLT agrees very well with the simulation results.

6.6 Aspects of SOQLT

The most important aspects of SOQLT are

- We can obtain SOQLT from QLT by applying the formal substitution of the resonance functions

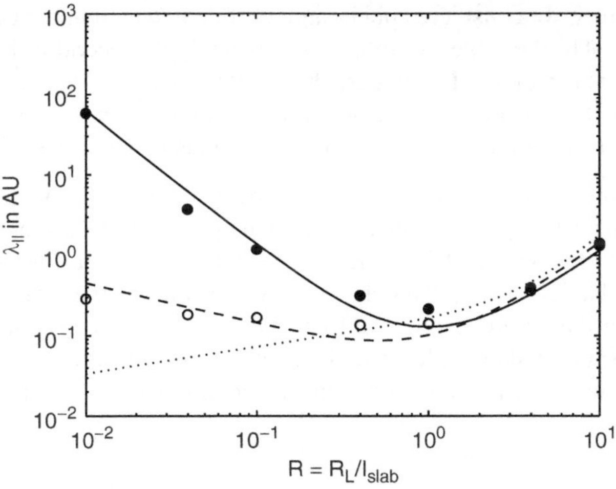

Fig. 6.6 $\lambda_{\|}$ for pure slab geometry and steep wave spectra. Simulations have been performed for $s = 2.5$ (*circles*) and $s = 2 + 5/3$ (*dots*). The simulations are compared with SOQLT results for $s = 2.5$ (*dashed line*) and $s = 2 + 5/3$ (*solid line*). The QLT result for such spectra is $\lambda_{\|} = \infty$. For comparison also the standard QLT result for $s = 5/3$ is shown (*dotted line*). In all cases SOQLT agrees very well with test-particle simulations

$$\begin{aligned} R_n^{\mathrm{QLT}} &= \pi\delta(k_{\|}v\mu + n\Omega) \\ \rightarrow R_n^{\mathrm{SOQLT}} &= \int_0^\infty \mathrm{d}t\, \cos\left[\left(k_{\|}v\mu + n\Omega\right)t\right] \mathrm{e}^{-v^2 k_{\|}^2 \int_0^t \mathrm{d}\tau \left[\tau^2 D_{\mu\mu}^{(1)}(\tau)\right]} \end{aligned} \tag{6.67}$$

in quasilinear formulas for $D_{\mu\mu}$.

- By using mathematical approximations (90°-approximation and 90°-LTA), simple analytical formulas for $D_{\mu\mu}(\mu = 0)$ can be derived. A numerical investigation of $D_{\mu\mu}$ for all values of μ as well as of $\lambda_{\|}$ as a function of rigidity is straightforward. Even if these approximations are applied, a reasonable agreement with simulations can be achieved.
- All analytical and numerical results presented in this chapter employed at least the 90°-approximation. Within the 90°-LTA, for instance, we find at 90° the formula

$$\tilde{D}_{\mu\mu}^{(2)}(\mu = 0) \approx \sqrt{\pi}\,\Gamma(\nu)C(\nu)R^{2\nu-2}\left(\frac{\delta B}{B_0}\right)^{2\nu+1}. \tag{6.68}$$

An improved analytical and numerical treatment without such approximations must be the subject of future work.

- SOQLT has been formulated only for slab geometry. To achieve a unified theory that can solve all three problems of cosmic ray transport theory, the ideas of SOQLT should be combined with WNLT.

- In this chapter, we considered pitch-angle diffusion only in magnetostatic turbulence. It must be the subject of future work to apply the second-order formulation to dynamical turbulence to compare 90°-scattering effects that arised from nonlinear corrections with the effect caused by dynamical turbulence. A recovery of QLT could be possible for such improved turbulence models (see Chap. 8).

By applying WNLT or SOQLT, we can describe pitch-angle diffusion at 90°, parallel transport in the composite model, and perpendicular diffusion in nonslab models. However, it is demonstrated in Chaps. 4 and 5 that, within all theories, we find a diffusive behavior of perpendicular transport in the magnetostatic slab model. Therefore, a further improvement of diffusion theory is necessary to understand subdiffusive transport in slab models and the recovery of diffusion in nonslab models. Such a theoretical investigation will be presented in the next chapter.

Chapter 7
The Extended Nonlinear Guiding Center Theory

In Chap. 4, the nonlinear guiding center theory of cosmic ray perpendicular diffusion was discussed. As in other linear and nonlinear approaches, we find a diffusive behavior of perpendicular transport for the magnetostatic slab model. This is in disagreement with the results of test-particle simulations, where subdiffusion is found. An extended theory, which includes an improved treatment of the slab contribution, has been developed by Shalchi (2006b) and is presented in this chapter. In contrast to the original theory, the extended nonlinear guiding center theory shows agreement with numerical simulations for slab and nonslab models.

7.1 The Slab Problem of Perpendicular Transport

In Chaps. 4 and 5, two promising nonlinear theories have been discussed: the nonlinear guiding center (NLGC) theory and the weakly nonlinear theory (WNLT). If we consider the NLGC theory, for instance, we find for pure slab geometry and magnetostatic turbulence (see (4.49))

$$\kappa_{xx} = 2\pi \frac{a^2 v^2}{3B_0^2} \int_{-\infty}^{+\infty} dk_\parallel \, g^{\text{slab}}(k_\parallel) \frac{1}{v/\lambda_\parallel + \kappa_{zz} k_\parallel^2}. \tag{7.1}$$

Obviously, within NLGC theory we find $\kappa_{xx} \neq 0$, which has to be interpreted as diffusion. In Chaps. 3 and 4, several theoretical approaches for perpendicular transport have been discussed. In Table 7.1, the results from these different theories are compared for the magnetostatic slab model. In these theories as well as within WNLT, we find a diffusive behavior of perpendicular transport. As shown and discussed in Chap. 3, we find subdiffusion of the form

$$\left\langle (\Delta x)^2 \right\rangle \sim t^{1/2} \tag{7.2}$$

in test-particle simulations (see Qin et al. 2002a). This subdiffusive behavior is in agreement with the Jokipii et al. (1993) theorem on reduced dimensionality, where it was argued that the perpendicular mean free path for pure slab turbulence must be

A. Shalchi, *Nonlinear Cosmic Ray Diffusion Theories*,
Astrophysics and Space Science Library,
DOI: 10.1007/978-3-642-00309-7_7,

Table 7.1 Different theoretical results for the perpendicular mean free path $\lambda_\perp$, obtained for magnetostatic slab turbulence

Theory	Case	$\lambda_\perp$
QLT	–	$\frac{3\pi}{2}C(\nu)l_{\mathrm{slab}}\left(\frac{\delta B}{B_0}\right)^2$
BAM model	$\Omega\tau \ll 1$	$\frac{2R^2}{3\pi C(\nu)}l_{\mathrm{slab}}\left(\frac{B_0}{\delta B}\right)^2$
	$\Omega\tau \gg 1$	$\frac{3\pi}{2}C(\nu)l_{\mathrm{slab}}\left(\frac{\delta B}{B_0}\right)^2$
NLGC theory	$\lambda_\parallel \gg \sqrt{3}l_{\mathrm{slab}}$	$\sqrt{3}\pi C(\nu)l_{\mathrm{slab}}\left(\frac{\delta B}{B_0}\right)^2$
	$\lambda_\parallel \ll \sqrt{3}l_{\mathrm{slab}}$	$\frac{1}{2}\lambda_\parallel\left(\frac{\delta B}{B_0}\right)^2$
NCA	$\lambda_\parallel/l_{\mathrm{slab}} \to 0$	$\frac{3\pi}{2}C(\nu)l_{\mathrm{slab}}\left(\frac{\delta B}{B_0}\right)^2$
	$\lambda_\parallel/l_{\mathrm{slab}} \to \infty$	$\frac{1}{2}\lambda_\parallel\left(\frac{\delta B}{B_0}\right)^2$
Simulations	–	Subdiffusion $\Rightarrow \lambda_\perp = 0$

Here, quasilinear theory is compared with the BAM model, the NLGC theory, the nonlinear closure approximation (NCA), and test-particle simulations (Qin et al. 2002a)

zero because particles are tied to magnetic lines of force. Furthermore, the calculations of Kóta and Jokipii (2000), based on the TGK formulation, provide the correct subdiffusive behavior according to (7.2). The aim of this chapter is to provide a systematic derivation of this result.

7.2 Integration of the Equations of Motion

By following Shalchi (2005b) we start with the assumption that guiding centers follow magnetic field lines (e.g., (3.4)). For pure slab turbulence, we have $\delta B(\mathbf{x}) = \delta B(z)$ and thus

$$\tilde{v}_x \equiv \frac{\mathrm{d}x}{\mathrm{d}t} = v_z \frac{\delta B_x}{B_0} = \frac{1}{B_0}\frac{\mathrm{d}}{\mathrm{d}t}\int \mathrm{d}^3k\, \frac{\delta B_x(k_\parallel)}{\mathrm{i}k_\parallel}\,\mathrm{e}^{\mathrm{i}k_\parallel z(t)}. \tag{7.3}$$

This equation can be integrated and we deduce

$$\Delta x(t) = \frac{1}{B_0}\int \mathrm{d}^3k\, \frac{\delta B_x(k_\parallel)}{\mathrm{i}k_\parallel}\left[\mathrm{e}^{\mathrm{i}k_\parallel z(t)} - \mathrm{e}^{\mathrm{i}k_\parallel z(0)}\right]. \tag{7.4}$$

For the mean square displacement, we therefore find

$$\begin{aligned}\left\langle(\Delta x)^2\right\rangle &= \frac{1}{B_0^2}\int \mathrm{d}^3k \int \mathrm{d}^3k'\, \frac{1}{k_\parallel k'_\parallel}\left\langle \delta B_x(k_\parallel)\delta B_x^*(k'_\parallel)\left[2 - \mathrm{e}^{+\mathrm{i}k_\parallel \Delta z(t)} - \mathrm{e}^{-\mathrm{i}k_\parallel \Delta z(t)}\right]\right\rangle,\\ &= \frac{1}{B_0^2}\int \mathrm{d}^3k\, \frac{P_{xx}(\mathbf{k})}{k_\parallel^2}\Gamma(k_\parallel, t),\end{aligned} \tag{7.5}$$

with

$$\Gamma(k_\parallel, t) = 2 - 2\left\langle \cos(k_\parallel \Delta z(t)) \right\rangle, \tag{7.6}$$

where again the Corrsin approximation (see (4.41)) has been used. For pure slab turbulence we can replace $P_{xx}(\mathbf{k})$ by (2.32) and we obtain

$$\left\langle (\Delta x)^2 \right\rangle = \frac{2\pi}{B_0^2} \int_{-\infty}^{+\infty} dk_\parallel \; g^{\text{slab}}(k_\parallel) k_\parallel^{-2} \Gamma(k_\parallel, t). \tag{7.7}$$

Obviously, we find a direct relation between the parallel motion of the particle (described by $\Gamma(k_\parallel, t)$) and the perpendicular motion (described by $\left\langle (\Delta x)^2 \right\rangle$). To proceed, we have to model the function $\Gamma(k_\parallel, t)$.

7.3 Application of Quasilinear Theory

Within QLT we replace the parallel position by the unperturbed orbit $z = v\mu t$, and the operator $\langle \dots \rangle$ in (7.6) can be dropped. In this case, the quasilinear approach means that we suppress parallel diffusion. For axisymmetric turbulence, the time dependent Fokker–Planck coefficient of perpendicular transport can be defined as $D_\perp(t) := \left\langle (\Delta x)^2 \right\rangle / (2t)$ and it follows

$$D_\perp(t) = \frac{2\pi v^2 \mu^2}{B_0^2} \int_{-\infty}^{+\infty} dk_\parallel \; g^{\text{slab}}(k_\parallel) K_0(t), \tag{7.8}$$

with

$$K_0(t) = \frac{1 - \cos(\beta_0 t)}{\beta_0^2 t} \tag{7.9}$$

and $\beta_0 = v\mu k_\parallel$. The usual time-independent Fokker–Planck coefficient can be calculated from

$$D_\perp = \lim_{t \to \infty} D_\perp(t). \tag{7.10}$$

By using (see e.g., Jaekel and Schlickeiser 1992)

$$\lim_{t \to \infty} \frac{1 - \cos(\beta_0 t)}{\beta_0^2 t} = \pi \delta(\beta_0), \tag{7.11}$$

we find

$$D_\perp = \frac{2\pi^2 v^2 \mu^2}{B_0^2} \int_{-\infty}^{\infty} dk_\parallel \; g^{\text{slab}}(k_\parallel) \delta(k v \mu) = \frac{2\pi^2 v}{B_0^2} |\mu| g^{\text{slab}}(0). \tag{7.12}$$

The perpendicular mean free path $\lambda_\perp$ can be obtained by considering the pitch-angle average of the Fokker–Planck coefficient $D_\perp$

$$\lambda_\perp = \frac{3}{2v}\int_{-1}^{+1} \mathrm{d}\mu\ D_\perp = \frac{3\pi^2}{B_0^2} g^{\mathrm{slab}}(0). \tag{7.13}$$

With the wave spectrum from (2.33), we find

$$\lambda_\perp = \frac{3\pi}{2} C(\nu) l_{\mathrm{slab}} \frac{\delta B_{\mathrm{slab}}^2}{B_0^2} =: \lambda_{\perp,\mathrm{FLRW}}, \tag{7.14}$$

which is the QLT result for perpendicular transport in the magnetostatic slab model (FLRW limit), originally proposed by Jokipii (1966) and also derived in Chap. 3 (see (3.48)).

7.3.1 Time-dependent Perpendicular Transport

Without taking the limit $t \to \infty$, the time-dependent perpendicular Fokker–Planck coefficient can be obtained from (7.8) and, for the dimensionless coefficient $\tilde{D}_\perp = D_\perp (v l_{\mathrm{slab}})^{-1}$, we find

$$\tilde{D}_\perp(t) = \frac{2\pi v \mu^2}{l_{\mathrm{slab}} B_0^2} \int_{-\infty}^{+\infty} \mathrm{d}k_\parallel\ g^{\mathrm{slab}}(k_\parallel) K_0(t). \tag{7.15}$$

With the dimensionless time $\tau := vt/l_{\mathrm{slab}}$, by using the integral transformation $x = k_\parallel l_{\mathrm{slab}}$ and by using the wave spectrum of (2.33), we have

$$\tilde{D}_\perp(\tau) = \frac{2C(\nu)\delta B_{\mathrm{slab}}^2}{B_0^2} \int_0^\infty \mathrm{d}x\ (1+x^2)^{-\nu} \frac{1-\cos(\mu x \tau)}{\tau x^2}. \tag{7.16}$$

Thus the time-dependent perpendicular mean free path $\lambda_\perp(\tau)$ has the form

$$\lambda_\perp(\tau) = \frac{6C(\nu) l_{\mathrm{slab}} \delta B_{\mathrm{slab}}^2}{B_0^2 \tau^2} \int_0^\infty \mathrm{d}x (1+x^2)^{-\nu} \frac{x\tau - \sin(x\tau)}{x^3}. \tag{7.17}$$

Figure 7.1 illustrates the numerically calculated $\lambda_\perp(\tau)$. To obtain the shown results, the following set of parameters

$$\begin{aligned} 2\nu &= 5/3, \\ l_{\mathrm{slab}} &= 0.03\,\mathrm{AU}, \\ R = R_{\mathrm{L}}/l_{\mathrm{slab}} &= 0.134, \\ \delta B/B_0 &= 1 \end{aligned} \tag{7.18}$$

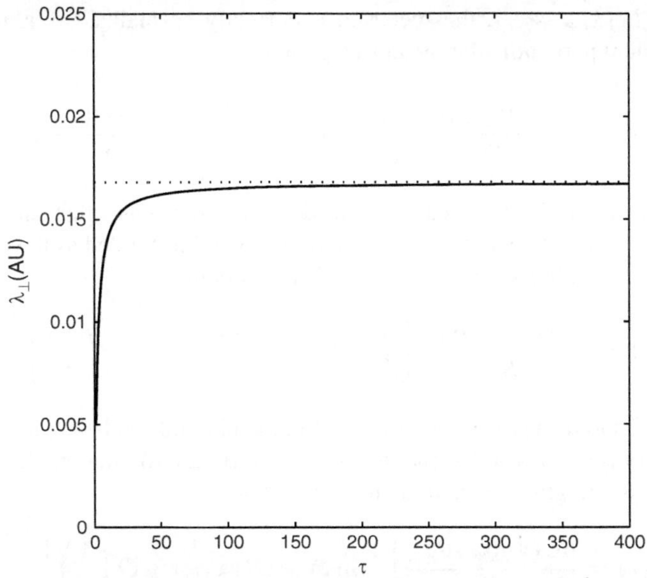

Fig. 7.1 The time-dependent perpendicular mean free path (*solid line*) and the diffusion limit (FLRW limit, *dotted line*). Both results were calculated by using quasilinear theory

has been used. After a certain time-scale, the perpendicular mean free path reaches the FLRW limit (7.14). Such a diffusive behavior disagrees with test-particle simulations, where subdiffusion was discovered.

7.3.2 *Finite Box-size Effects*

Equation (7.14) is the quasilinear result for perpendicular diffusion. Real physical systems as well as the simulated systems have a finitely large size L (box-size) and, therefore, the existence of a minimal wavenumber $k_{\min} \approx L^{-1}$ is obvious. Thus, it seems more appropriate to assume $g^{\text{slab}}(0) = 0$ and, with (7.13), one would find $\lambda_\perp = 0$, which may be interpreted as subdiffusive behavior. In the following, it is demonstrated that the FLRW limit is indeed the correct QLT result even if we take into account the limited system size. For the following discussions, we assume a wave spectrum of the form

$$g^{\text{slab}}(k_\parallel) = \frac{C(\nu)}{2\pi} l_{\text{slab}} \delta B_{\text{slab}}^2 \begin{cases} 0 & \text{for } k_\parallel < k_{\min} \\ (1 + k_\parallel^2 l_{\text{slab}}^2)^{-\nu} & \text{for } k_\parallel \geq k_{\min}. \end{cases} \tag{7.19}$$

As long as $l_{\text{slab}} k_{\text{min}} \ll 1$, the spectrum is correctly normalized. Within QLT, the time-dependent perpendicular mean free path is

$$\lambda_\perp(\tau) = \frac{6C(\nu) l_{\text{slab}} \delta B_{\text{slab}}^2}{B_0^2} \int_{x_{\text{min}}}^{\infty} \mathrm{d}x \; (1+x^2)^{-\nu} \frac{x\tau - \sin(x\tau)}{\tau^2 x^3}, \tag{7.20}$$

with $x_{\text{min}} = l_{\text{slab}}/L$. To proceed, we consider only time-scales that satisfy the condition $l_{\text{slab}} \ll vt \ll L$. The second restriction can be interpreted as the existence of a maximal time scale $t \ll L/v =: t_{\text{max}}$. We then obtain

$$\lambda_\perp(\tau) = \frac{6C(\nu) l_{\text{slab}} \delta B_{\text{slab}}^2}{B_0^2} \left[\int_{x_{\text{min}}}^{1} \mathrm{d}x \; \frac{x\tau - \sin(x\tau)}{x^3 \tau^2} + \mathcal{O}\left(\frac{1}{\tau}\right) \right]. \tag{7.21}$$

According to Gradshteyn and Ryzhik (2000), the integral can be expressed in terms of elementary functions and a sine integral $\text{Si}(x)$. By approximating the elementary functions, it is straightforward to demonstrate that

$$\lambda_\perp(\tau) = \frac{6C(\nu) l_{\text{slab}} \delta B_{\text{slab}}^2}{B_0^2} \left[\frac{1}{2} \text{Si}(\tau) + \mathcal{O}(\tau x_{\text{min}}) + \mathcal{O}\left(\frac{1}{\tau}\right) \right]. \tag{7.22}$$

Using $\text{Si}(\tau \gg 1) \approx \pi/2$ (Abramowitz and Stegun 1974), we finally find $\lambda_\perp(\tau) = \lambda_{\perp,\text{FLRW}}$. It seems that if we restrict not only the length scale but also the time scale, we still obtain the FLRW limit. The restriction that must be applied to obtain the FLRW limit is $vt \le L$. For larger time-scales, the particle would have left

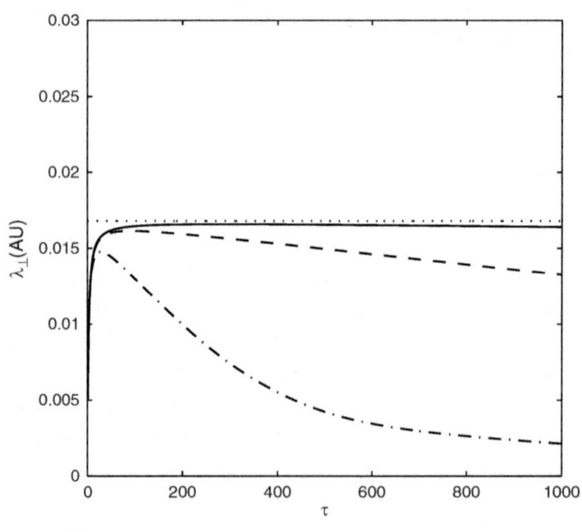

Fig. 7.2 The time-dependent perpendicular mean free path for different values of the minimal wavenumber: $x_{\text{min}} = 10^{-4}$ (*solid line*), $x_{\text{min}} = 10^{-3}$ (*dashed line*), and $x_{\text{min}} = 10^{-2}$ (*dash-dotted line*). Also, the FLRW limit is shown (*dotted line*). All results are calculated by using quasilinear theory

the physical system and the diffusion coefficient tends to zero. Figure 7.2 shows numerical solutions of (7.20) for different values of $x_{\rm min}$, which confirms this result: for times larger than $L/v \sim x_{\rm min}^{-1}$, perpendicular transport would become subdiffusive, but such time-scales are forbidden. For the values used in test-particle simulations ($x_{\rm min} \ll 10^{-1}$), we always find diffusion for the interesting time-scales (see Fig. 7.2).

7.4 The Nonlinear Guiding Center Model

The results of the previous section do not agree with simulations, where subdiffusion has been discovered. Therefore, the quasilinear approach that has been applied so far is not acceptable. The reason for this disagreement can easily be understood: in (7.7), we found a direct relation between the parallel and the perpendicular motion of the particle. For the considered time scales, for which we expect a stable motion (e.g., a subdiffusive motion in the considered case) of the particle in the perpendicular direction, we know that parallel transport behaves diffusively. Therefore, it is insentient to assume unperturbed orbits. In the following, a different treatment of (7.6) is used to explain subdiffusion.

By assuming a Gaussian form of the particle distribution function and assuming that parallel scattering behaves diffusively, we can apply the model

$$\left\langle e^{\pm i k_\parallel z} \right\rangle = e^{-\kappa_\parallel k_\parallel^2 t}, \tag{7.23}$$

which has also been used during the derivation of NLGC theory (see (4.44)). Then the function $\Gamma(k_\parallel, t)$ of (7.6) becomes

$$\Gamma(k_\parallel, t) = 2 - 2\, e^{-\kappa_\parallel k_\parallel^2 t}, \tag{7.24}$$

which can be combined with (7.5) to obtain

$$\left\langle (\Delta x)^2 \right\rangle = \frac{8\pi}{B_0^2} \int_0^\infty dk_\parallel \; g^{\rm slab}(k_\parallel) k_\parallel^{-2} \left[1 - e^{-\kappa_\parallel k_\parallel^2 t} \right]. \tag{7.25}$$

For the spectrum from (2.33), we find

$$\begin{aligned} \left\langle (\Delta x)^2 \right\rangle = {} & 8 C(\nu) l_{\rm slab} \left(\frac{\delta B_{\rm slab}^2}{B_0^2} \right)^2 \\ & \times \int_{k_{\rm min}}^\infty dk_\parallel \; (1 + k_\parallel^2 l_{\rm slab})^{-\nu} k_\parallel^{-2} \left[1 - e^{-\kappa_\parallel k_\parallel^2 t} \right]. \end{aligned} \tag{7.26}$$

Again, we calculate the time-dependent perpendicular mean free path

$$\lambda_\perp = \frac{3}{v}\kappa_\perp = \frac{3}{2tv}\left\langle (\Delta x)^2 \right\rangle. \tag{7.27}$$

With

$$\gamma := \sqrt{\frac{\lambda_\| \tau}{3 l_{\text{slab}}}}, \tag{7.28}$$

$x = k_\| l_{\text{slab}}$, and $\tau = vt / l_{\text{slab}}$, one can deduce

$$\lambda_\perp(\tau) = \frac{6C(\nu) l_{\text{slab}} \delta B_{\text{slab}}^2}{\tau B_0^2} \int_{x_{\min}}^{\infty} \mathrm{d}x \, (1+x^2)^{-\nu} x^{-2} \left[1 - \mathrm{e}^{-\gamma^2 x^2}\right]. \tag{7.29}$$

In the following, this result is investigated analytically and numerically.

7.4.1 Analytical and Numerical Results of the Nonlinear Model

To consider analytical results of (7.29), we restrict our calculations to $x_{\min} \ll \gamma^{-1} \ll 1$, which can be written as

$$\tau_{\max} := \frac{3 l_{\text{slab}}}{\lambda_\| x_{\min}^2} \gg \tau \gg \frac{3 l_{\text{slab}}}{\lambda_\|} =: \tau_{\min}, \tag{7.30}$$

where the minimal time-scale $\tau_{\min}$ and the maximal time-scale $\tau_{\max}$ have been defined. Now we split the integral of (7.29) to get

$$\lambda_\perp(\tau) \approx \frac{6C(\nu) l_{\text{slab}} \delta B_{\text{slab}}^2}{\tau B_0^2} \left(\int_{x_{\min}}^{1} \mathrm{d}x \, \frac{1 - \mathrm{e}^{-\gamma^2 x^2}}{x^2} + \int_{1}^{\infty} \mathrm{d}x \, \frac{1 - \mathrm{e}^{-\gamma^2 x^2}}{x^{2+2\nu}} \right). \tag{7.31}$$

The first integral can be expressed in terms of elementary functions and error functions (see Gradshteyn and Ryzhik 2000)

$$\begin{aligned} \int_{x_{\min}}^{1} \mathrm{d}x \, \frac{1 - \mathrm{e}^{-\gamma^2 x^2}}{x^2} &= \mathrm{e}^{-\gamma^2} - 1 + x_{\min}^{-1} \left(1 - \mathrm{e}^{-\gamma^2 x_{\min}^2}\right) \\ &\quad + \gamma\sqrt{\pi}\,\mathrm{erf}(\gamma) - \gamma\sqrt{\pi}\,\mathrm{erf}(\gamma x_{\min}) \end{aligned} \tag{7.32}$$

and the second integral behaves like τ^{-1} and can, therefore, be neglected. For the condition of (7.30), the error functions can be approximated (see Abramowitz and Stegun 1974)

$$\operatorname{erf}(z \ll 1) \approx \frac{2z}{\sqrt{\pi}},$$
$$\operatorname{erf}(z \gg 1) \approx 1, \tag{7.33}$$

and one gets approximately

$$\int_{x_{\min}}^{1} \mathrm{d}x \, \frac{1 - \mathrm{e}^{-\gamma^2 x^2}}{x^2} = \sqrt{\pi}\gamma. \tag{7.34}$$

Finally, we find for the time-dependent perpendicular mean free path

$$\lambda_{\perp}(\tau) \approx 4\lambda_{\perp,\mathrm{FLRW}} \sqrt{\frac{\lambda_{\parallel}}{3\pi l_{slab}}}\, \tau^{-1/2}. \tag{7.35}$$

Like in the quasilinear approach, we can also understand the restriction of (7.30): the width of the Gaussian particle distribution function cannot exceed the box-size. Obviously, we find for the interesting time-scales $\langle(\Delta x)^2\rangle \sim \sqrt{t}$ in agreement with the test-particle simulations. Figure 7.3 shows the nonlinear $\lambda_{\perp}(\tau)$ for $x_{\min} = 10^{-4}$. For late times, we find $\lambda_{\perp}(\tau) \sim \tau^{-1/2}$, which is in agreement with simulations (e.g., Qin et al. 2002a, b), with previous theoretical discussions (Kóta and Jokipii

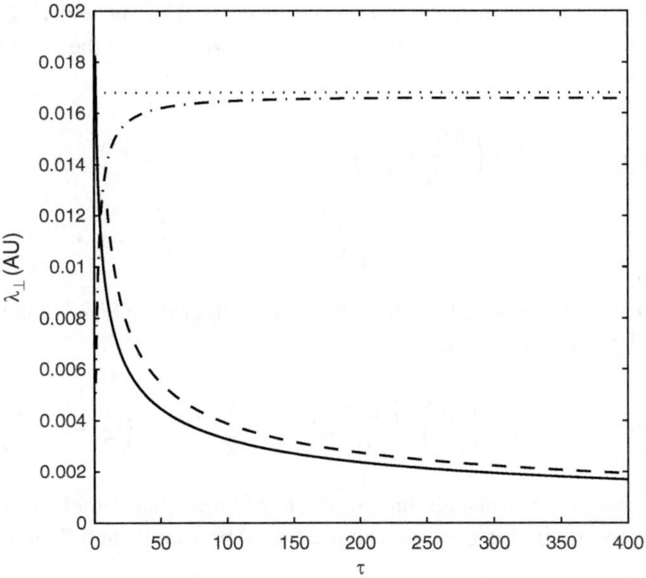

Fig. 7.3 The time-dependent nonlinear perpendicular mean free path (*solid line*) and the FLRW limit (*dotted line*). The nonlinear diffusion coefficient shows subdiffusive behavior ($\lambda_{\perp} \sim \tau^{-1/2}$). The *dashed line* is the analytical result of (7.35), which can only be applied for $\tau \gg 1$, and the *dash-dotted line* is the time-dependent quasilinear result, which behaves like $\lambda_{\perp} \sim const$

2000) and with the analytical results of this book (7.35). A direct comparison with simulations will be presented in Sect. 7.6.

7.4.2 *Running Diffusion Coefficient and Velocity Correlation Function*

Besides the time-dependent perpendicular spatial diffusion coefficient (or time-dependent perpendicular mean free path), two other interesting transport parameters can be calculated: the running diffusion coefficient $d_{xx}(t)$ defined in (1.21) and the velocity correlation function $V_{xx}(t)$. By using the definition of a running diffusion coefficient (1.21), we find from (7.26)

$$d_{xx}(t) \equiv \frac{1}{2}\frac{\mathrm{d}}{\mathrm{d}t}\left\langle (\Delta x)^2 \right\rangle = \frac{4\pi\kappa_\parallel}{B_0^2}\int_0^\infty \mathrm{d}k_\parallel\, g^{\mathrm{slab}}(k_\parallel)\, \mathrm{e}^{-\kappa_\parallel k_\parallel^2 t}. \tag{7.36}$$

From this result, an analytical form for the velocity correlation function can be derived by applying (1.30)

$$V_{xx}(t) \equiv \frac{\mathrm{d}}{\mathrm{d}t} d_{xx} = -\frac{4\pi\kappa_\parallel^2}{B_0^2}\int_0^\infty \mathrm{d}k_\parallel\, g^{\mathrm{slab}}(k_\parallel) k_\parallel^2\, \mathrm{e}^{-\kappa_\parallel k_\parallel^2 t}. \tag{7.37}$$

To proceed, we use again the standard spectrum (2.33). By applying the integral transformation $x = k_\parallel l_{\mathrm{slab}}$ and by using (7.28), we find for the running diffusion coefficient

$$\begin{aligned} d_{xx}(t) &= 2C(\nu)\kappa_\parallel \left(\frac{\delta B_{\mathrm{slab}}}{B_0}\right)^2 \int_0^\infty \mathrm{d}x\, (1+x^2)^{-\nu}\, \mathrm{e}^{-\gamma^2 x^2} \\ &\approx 2C(\nu)\kappa_\parallel \left(\frac{\delta B_{\mathrm{slab}}}{B_0}\right)^2 \left[\int_0^1 \mathrm{d}x\, \mathrm{e}^{-\gamma^2 x^2} + \int_1^\infty \mathrm{d}x\, x^{-2\nu}\, \mathrm{e}^{-\gamma^2 x^2}\right]. \end{aligned} \tag{7.38}$$

According to Gradshteyn and Ryzhik (2000), the integrals in (7.38) can be expressed by error and gamma functions:

$$d_{xx}(t) = 2C(\nu)\kappa_\parallel \left(\frac{\delta B_{\mathrm{slab}}}{B_0}\right)^2 \left[\frac{\sqrt{\pi}}{2\gamma}\mathrm{erf}(\gamma) + \frac{1}{2}\gamma^{2\nu-1}\Gamma\left(\frac{1}{2}-\nu, \gamma^2\right)\right]. \tag{7.39}$$

In the following, we consider the limit of infinitely late times ($t \to \infty$) and, therefore, $\gamma \to \infty$. In this case we can use (see Abramovitz and Stegun 1974)

$$\begin{aligned} \mathrm{erf}(\gamma) &\to 1, \\ \Gamma\left(\frac{1}{2}-\nu, \gamma^2\right) &\to \gamma^{-2\nu-1}\, \mathrm{e}^{-\gamma^2} \end{aligned} \tag{7.40}$$

to obtain

$$d_{xx}(t \to \infty) \approx C(\nu) l_{\text{slab}} \left(\frac{\delta B_{\text{slab}}}{B_0} \right)^2 \sqrt{\frac{\pi \kappa_\parallel}{t}}. \tag{7.41}$$

By applying (7.37), we find for the perpendicular velocity correlation function

$$V_{xx}(t \to \infty) = -\frac{\sqrt{\pi}}{2} C(\nu) l_{\text{slab}} \sqrt{\kappa_\parallel} \left(\frac{\delta B_{\text{slab}}}{B_0} \right)^2 t^{-3/2}. \tag{7.42}$$

Obviously, the velocity correlation function has a nonexponential tail. This non-exponential behavior of velocity correlation functions is the reason why previous nonlinear theories failed to reproduce the simulations.

7.5 The Extended Nonlinear Guiding Center Theory

In the previous section, an improved or extended approach has been discussed to describe perpendicular transport in the magnetostatic slab model. This approach is more reliable than the previous theories such as NLGC theory, BAM model, or WNLT. To describe perpendicular transport in the slab/2D composite model, the calculations of the previous section are combined with the standard NLGC theory. Again, the starting point of our calculations is the assumption that guiding centers follow magnetic field lines:

$$v_x = v_z \frac{\delta B_x(x, y, z)}{B_0} = v_z \frac{\delta B_x^{\text{slab}}(z)}{B_0} + v_z \frac{\delta B_x^{\text{2D}}(x, y)}{B_0}. \tag{7.43}$$

By applying a Fourier transformation, (7.43) can be rewritten as

$$\frac{\mathrm{d}}{\mathrm{d}t} x = \frac{\mathrm{d}}{\mathrm{d}t} \int \mathrm{d}^3 k \, \frac{\delta B_x^{\text{slab}}(k_\parallel)}{\mathrm{i} k_\parallel B_0} \, \mathrm{e}^{\mathrm{i} k_\parallel z} + v_z \int \mathrm{d}^3 k \, \frac{\delta B_x^{\text{2D}}(k_x, k_y)}{B_0} \, \mathrm{e}^{\mathrm{i} k_x x + \mathrm{i} k_y y} \tag{7.44}$$

and thus

$$\begin{aligned} \Delta x(t) = & \int \mathrm{d}^3 k \, \frac{\delta B_x^{\text{slab}}(k_\parallel)}{\mathrm{i} k_\parallel B_0} \left[\mathrm{e}^{\mathrm{i} k_\parallel z(t)} - 1 \right] \\ & + \int_0^t \mathrm{d}\tau \, v_z(\tau) \int \mathrm{d}^3 k \, \frac{\delta B_x^{\text{2D}}(k_x, k_y)}{B_0} \, \mathrm{e}^{\mathrm{i} k_x x + \mathrm{i} k_y y}, \end{aligned} \tag{7.45}$$

where we have used $\Delta x = x(t) - x(0)$ and $z(0) = 0$. By employing $\kappa_{xx}(t) = \langle (\Delta x)^2 \rangle / (2t)$ and Corrsin's independence hypothesis (4.41), we find

$$\begin{aligned}\kappa_{xx}(t) &= \frac{1}{2tB_0^2}\int d^3k\, P_{xx}^{\mathrm{slab}}(\mathbf{k})\frac{2-2\langle\cos\left[k_\parallel z(t)\right]\rangle}{k_\parallel^2}\\&+\frac{1}{2tB_0^2}\int_0^t d\tau_1\int_0^t d\tau_2\,\langle v_z(\tau_1)v_z(\tau_2)\rangle\\&\times\int d^3k\, P_{xx}^{\mathrm{2D}}(\mathbf{k})\left\langle e^{ik_x[x(\tau_1)-x(\tau_2)]+ik_y[y(\tau_1)-y(\tau_2)]}\right\rangle.\end{aligned} \tag{7.46}$$

The slab contribution can be calculated in the same manner as described in Sect. 7.4. For $P_{xx}^{\mathrm{slab}}(\mathbf{k})$, we use again (2.32) with (2.33), and we therefore obtain

$$\lim_{t\to\infty}\kappa_{xx}^{\mathrm{slab}}(t) \approx 2\sqrt{\pi}C(\nu)l_{\mathrm{slab}}\frac{\delta B_{\mathrm{slab}}^2}{B_0^2}\sqrt{\frac{\kappa_\parallel}{t}}. \tag{7.47}$$

Obviously, the slab contribution also behaves subdiffusively in the composite model. The 2D contribution is much more difficult to calculate. The main problem is that the time-integrals cover many scales, from small time-scales, where we expect a ballistic motion of the particle, to large, diffusive scales. It is difficult to take into account these different time-scales if a nonlinear theory is formulated. Here, we apply the models and approximations suggested by Matthaeus et al. (2003) (see Chap. 4):

1. We assume a Gaussian distribution of the particles (or guiding centers) for all time-scales
2. We assume that the mean square deviation of the particle (the width of the Gaussian function) behaves diffusively for all times:

$$\left\langle e^{ik_x[x(\tau_1)-x(\tau_2)]+ik_y[y(\tau_1)-y(\tau_2)]}\right\rangle = e^{-(\kappa_{xx}k_x^2+\kappa_{yy}k_y^2)|\tau_1-\tau_2|}. \tag{7.48}$$

   Furthermore, we neglect the gyromotion of the particle.
3. We assume an isotropic and exponential form of the velocity correlation function

$$\langle v_z(\tau_1)v_z(\tau_2)\rangle = \frac{v^2}{3}e^{-v|\tau_1-\tau_2|/\lambda_\parallel}. \tag{7.49}$$

By applying these three approximations, we can determine the 2D contribution:

$$\kappa_{xx}^{\mathrm{2D}}(t) = \frac{v^2}{6tB_0^2}\int d^3k\, P_{xx}^{\mathrm{2D}}(\mathbf{k})\int_0^t d\tau_1\int_0^t d\tau_2\, e^{-(\kappa_{xx}k_x^2+\kappa_{yy}k_y^2+v/\lambda_\parallel)|\tau_1-\tau_2|}. \tag{7.50}$$

Because we are interested in large (diffusive) time scales, we can calculate the two time integrals for the limit $t \to \infty$:

$$\lim_{t\to\infty} \kappa_{xx}^{2D}(t) = \frac{v^2}{3B_0^2} \int d^3k \, \frac{P_{xx}^{2D}(\mathbf{k})}{\kappa_{xx}k_x^2 + \kappa_{yy}k_y^2 + v/\lambda_\parallel}. \tag{7.51}$$

To replace $P_{xx}^{2D}(\mathbf{k})$, we use (2.50) with the spectrum of (2.51) to obtain

$$\lim_{t\to\infty} \kappa_{xx}^{2D}(t) = \frac{2v^2 C(\nu) l_{2D} \delta B_{2D}^2}{3B_0^2} \int_0^\infty dk_\perp \, \frac{(1 + k_\perp^2 l_{2D}^2)^{-\nu}}{\kappa_\perp k_\perp^2 + v/\lambda_\parallel}. \tag{7.52}$$

Equation (7.52) describes the same 2D contribution that was already calculated in Chap. 4. The total diffusion coefficient consists of a subdiffusive slab (7.47) and a diffusive 2D contribution (7.52).

7.5.1 *Analytic Forms of the Perpendicular Mean Free Path*

With astrophysical applications in mind, it is useful to derive analytic forms of the perpendicular mean free path. As shown in Shalchi et al. (2004a), the integral equation of (7.52) can be solved analytically:

$$\lambda_\perp = \frac{1}{2} \frac{\delta B_{2D}^2}{B_0^2} \lambda_\parallel \tag{7.53}$$

if $\lambda_\parallel \lambda_\perp \ll 3 l_{2D}^2$, and

$$\lambda_\perp = \left(\sqrt{3} \pi C(\nu) l_{2D} \frac{\delta B_{2D}^2}{B_0^2} \right)^{2/3} \lambda_\parallel^{1/3} \tag{7.54}$$

if $\lambda_\parallel \lambda_\perp \gg 3 l_{2D}^2$. These two formulas must be used if the perpendicular mean free path is calculated for composite or pure 2D geometry. For pure slab, we must employ (7.47). These formulas can easily be applied and are, therefore, useful for applications in space- and astrophysics.

7.6 Comparison with Test-particle Simulations

According to the new results discussed here, we can formulate the following statement: *the slab contribution always behaves subdiffusively and must, therefore, be neglected if the perpendicular mean free path is calculated for a magnetostatic composite model.* Here, we test this statement by comparing the original and the improved (extended) NLGC theory with test-particle simulations. Furthermore, it seems that there are three different time-regimes:

1. The small scales where the particle moves nearly unperturbed (ballistic regime)
2. The intermediate scales, at which the subdiffusive slab contribution is dominant. For such time scales the total mean free path behaves subdiffusively.
3. The large scales, where the time is large enough so that the slab contribution can be neglected and diffusion is recovered. In the case of pure slab geometry, this third regime does not exist.

To test the ENLGC theory, we compare with test-particle simulations. We perform three runs to explore pure slab, strong slab, and strong 2D turbulence. For the turbulence parameters, we use $l_{\text{slab}} = 10 l_{2\text{D}} = 0.03\,\text{AU}$, $2\nu = 5/3$, and $\delta B/B_0 = 1$ in the simulations and in the theoretical calculations. If the NLGC or ENLGC theory is applied, the parallel mean free path must be known as an input parameter. To replace the parallel mean free path, we use the results from simulations. Both diffusion coefficients are calculated for a value of the dimensionless rigidity $R = R_{\text{L}}/l_{\text{slab}} = 0.1$. For the NLGC results we employ a parameter value of $a^2 = 1$, which corresponds to the assumption that guiding centers follow magnetic field lines.

7.6.1 Run 1: Pure Slab Geometry

In this first run, we consider pure slab geometry. In this case, the ENLGC result is represented by (7.47), which can be written as

$$\lambda_\perp = 6C(\nu)\frac{\delta B_{\text{slab}}^2}{B_0^2}\sqrt{\frac{\pi\lambda_\parallel l_{\text{slab}}}{3\tau}}, \tag{7.55}$$

where $\lambda_i = 3/v\kappa_i$ and where the dimensionless time $\tau = vt/l_{\text{slab}}$ has been used. In Fig. 7.4, this formula is compared with the time-dependent result from the simulations. Both ENLGC and the simulations provide a subdiffusive behavior: $\lambda_\perp \sim \tau^{-1/2}$. The QLT and the NLGC results are constant in time and disagree with the simulations. To substitute the parallel mean free path in the theoretical results, the simulated values have been used.

7.6.2 Run 2: Strong Slab Geometry

In the second run, we assume 90% slab and 10% 2D geometry. In this case, diffusion is recovered or nearly recovered (see Fig. 7.5). To obtain the ENLGC result, we neglect the slab contribution. In this case, the original NLGC theory fails to

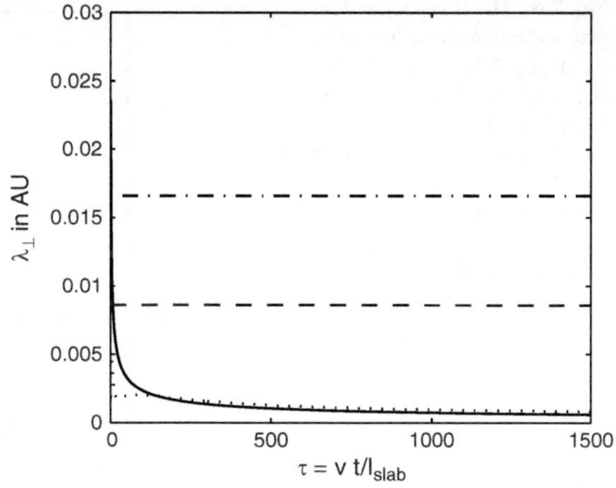

Fig. 7.4 The time-dependent perpendicular mean free path for pure slab geometry. Shown is the QLT result (*dash-dotted line*), the NLGC result (*dashed line*), the ENLGC result (*solid line*), and the simulations (*dotted line*)

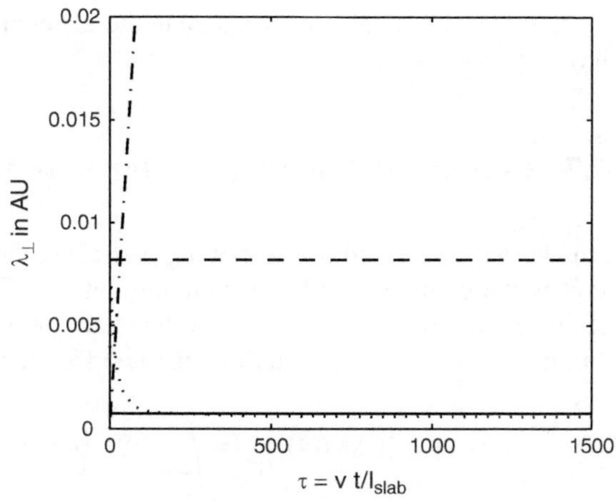

Fig. 7.5 The time-dependent perpendicular mean free path for strong slab geometry (90% slab/10% 2D). Shown is the NLGC result (*dashed line*), the ENLGC result (*solid line*), and the simulations (*dotted line*). The QLT result for the considered parameters is superdiffusive (*dash-dotted line*)

reproduce the simulations due to overestimating the slab contribution. The ENLGC theory agrees very well with the simulations. The statement regarding the slab contribution being neglected seems to be true. The QLT result for this parameter regime is $\lambda_\perp^{\text{QLT}} \sim t \to \infty$ due to the superdiffusive behavior (see Chap. 3).

7.6.3 Run 3: Strong 2D Geometry

In Fig. 7.6, the results for 20% slab/80% 2D composite geometry are illustrated. Here, the NLGC and the ENLGC results are similar. The factor 2 between the

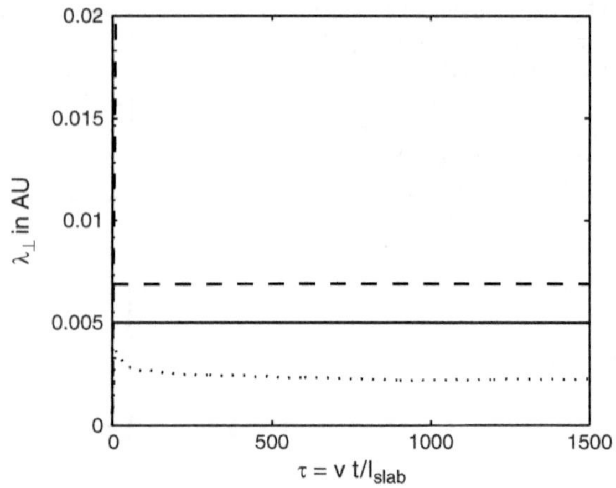

Fig. 7.6 The time-dependent perpendicular mean free path for strong 2D geometry (20% slab/80% 2D). Shown is the NLGC result (*dashed line*), the ENLGC result (*solid line*), and the simulations (*dotted line*). The QLT result for the considered parameters is, again, superdiffusive (*dash-dotted line*)

extended theory and the simulations is expected to be caused by the crude models that have been applied to calculate the 2D contribution. Again, within QLT we have $\lambda_\perp^{\mathrm{QLT}} \sim t \to \infty$.

7.7 Compound Subdiffusion for Pure Slab Turbulence

An alternative method for describing particle scattering in the perpendicular direction is the compound diffusion model that can be obtained from the Chapman–Kolmogorov approach (see, e.g., Webb et al. 2006). If we assume that the particles (or, more precisely, their guiding-centers) follow the magnetic field-lines, we can use

$$\left\langle (\Delta x(t))^2 \right\rangle_{\mathrm{P}} = \int_{-\infty}^{+\infty} \mathrm{d}z \left\langle (\Delta x(z))^2 \right\rangle_{\mathrm{FL}} f_{\mathrm{P}}(z,t). \tag{7.56}$$

Here $\langle (\Delta x(t))^2 \rangle_{\mathrm{P}}$ denotes the perpendicular MSD (mean square deviation) of the charged particle, $\langle (\Delta x(z))^2 \rangle_{\mathrm{FL}}$ is the field line MSD, and $f_{\mathrm{P}}(z,t)$ is the particle distribution in the direction parallel to the background field. By assuming again a Gaussian particle distribution

$$f_{\mathrm{P}}(z,t) = \left(2\pi \left\langle (\Delta z(t))^2 \right\rangle_{\mathrm{P}} \right)^{-1/2} \mathrm{e}^{-\frac{z^2}{2\left\langle (\Delta z(t))^2 \right\rangle_{\mathrm{P}}}} \tag{7.57}$$

and diffusive field line random walk (see Chap. 2)

$$\left\langle (\Delta x(z))^2 \right\rangle_{\mathrm{FL}} = 2 \mid z \mid \kappa_{\mathrm{FL}} \tag{7.58}$$

with the field line diffusion coefficient κ_{FL} (7.56) becomes

$$\left\langle (\Delta x(t))^2 \right\rangle_{\mathrm{P}} = \frac{4\kappa_{\mathrm{FL}}}{\sqrt{2\pi \left\langle (\Delta z(t))^2 \right\rangle_{\mathrm{P}}}} \int_0^{\infty} \mathrm{d}z\, z\, \mathrm{e}^{-\frac{z^2}{2\left\langle (\Delta z(t))^2 \right\rangle_{\mathrm{P}}}}. \tag{7.59}$$

By additionally assuming that parallel scattering is diffusive

$$\left\langle (\Delta z(t))^2 \right\rangle_{\mathrm{P}} = 2\kappa_{\parallel} t, \tag{7.60}$$

this becomes

$$\left\langle (\Delta x(t))^2 \right\rangle_{\mathrm{P}} = 4\kappa_{\mathrm{FL}} \sqrt{\frac{\kappa_{\parallel} t}{\pi}}. \tag{7.61}$$

For the running diffusion coefficient, we then obtain

$$\begin{aligned} d_{xx}(t) &\equiv \frac{1}{2} \frac{\mathrm{d}}{\mathrm{d}t} \left\langle (\Delta x(t))^2 \right\rangle_{\mathrm{P}} \\ &= \kappa_{\mathrm{FL}} \sqrt{\frac{\kappa_{\parallel}}{\pi t}}. \end{aligned} \tag{7.62}$$

With (2.89) for the field line diffusion coefficient κ_{FL}, we find

$$d_{xx} = C(\nu) l_{\mathrm{slab}} \frac{\delta B_{\mathrm{slab}}^2}{B_0^2} \sqrt{\frac{\pi \kappa_{\parallel}}{t}}. \tag{7.63}$$

This result is in agreement with the result derived by employing the ENLGC theory (see (7.41)) and the result derived by Kóta and Jokipii (2000). The interesting aspect of the compound diffusion model is that the theory is much more systematic than the other approaches. Please note that (7.61) and (7.63) are correct if the following requirements are fulfilled:

- Charged particles follow magnetic field lines.
- FLRW behaves diffusively with the field line diffusion coefficient κ_{FL}.
- The particles are distributed like a Gaussian function in the parallel direction.
- Parallel scattering behaves diffusively with the parallel diffusion coefficient $\kappa_{\parallel}$.

If these requirement are fulfilled, we always obtain a strong subdiffusive perpendicular scattering with

$$\left\langle (\Delta x(t))^2 \right\rangle_{\mathrm{P}} \sim \sqrt{t}. \tag{7.64}$$

For slab turbulence, this result seems to be correct. For other models, such as the slab/2D model, however, it is known that perpendicular scattering is diffusive or at least very close to the diffusive regime. Thus, we can conclude that for more realistic turbulence models, one of the four assumptions above has to be wrong. In Chap. 2, we have already demonstrated that FLRW behaves superdiffusively for the slab/2D model. The combination of the compound transport model with realistic turbulence models, for which we expect superdiffusive FLRW, and a detailed comparison with test-particle simulations, has to be the subject of future work. Some further aspects of the compound diffusion model are discussed in Shalchi et al. (2007).

7.8 Aspects of ENLGC Theory

The following results have been obtained in this chapter:

- It is demonstrated that the parallel motion of the particle determines the perpendicular motion at exactly the same time (see (7.7)) if the magnetostatic slab model is applied. Modeling the velocity correlation function as in NLGC theory or WNLT is not necessary.
- By applying QLT to replace the parallel motion of the particle and by applying a wave spectrum with small wavenumber cutoff, it is demonstrated that the FLRW limit (see (7.14)) is the correct quasilinear result for $\lambda_\perp$, even in cases where $g^{\text{slab}}(k_\parallel = 0) = 0$[1].
- By replacing the unperturbed motion of the particle by a diffusive motion, the subdiffusive behavior of perpendicular transport can be derived systematically. Obviously, parallel diffusion suppresses perpendicular transport to a subdiffusive level.
- By combining this improved treatment of the slab contribution with standard NLGC theory for the 2D contribution, an extended NLGC theory can be derived. This theory can quantitatively describe subdiffusion for slab and the recovery of diffusion for nonslab models. As demonstrated, the ENLGC theory is in agreement with test-particle simulations.
- For pure slab turbulence, the ENLGC result from (7.41) agrees with the result obtained from the compound diffusion model (see e.g., Kóta and Jokipii 2000; Webb et al. 2006; Shalchi et al. 2007; and (7.63) of this book). The strong subdiffusive behavior is also in agreement with the Jokipii et al. (1993) theorem on reduced dimensionality.
- The modifications that have been applied to the NLGC approach must also be applied to the weakly nonlinear theory. Within WNLT, we must set $D_\perp^{\text{slab}} \to 0$, where $D_\perp^{\text{slab}}$ is the slab contribution to the Fokker–Planck coefficient of perpendicular diffusion. Therefore, the following set of equations seems to be more appropriate than the set presented in Chap. 5 ((5.25), (5.31), (5.34), and (5.40)):

[1] This statement is based on the assumption that particles follow field lines. If the particles can be scattered away from a single field line, the linear result presented in this book could become inaccurate.

$$
\begin{aligned}
\tilde{D}_{\mu\mu}^{\text{slab}} &= \pi C(\nu)(1-\mu^2)|\mu|^{2\nu-1} R^{2\nu-2} \frac{\delta B_{\text{slab}}^2}{B_0^2} \left(\mu^2 R^2 + 1\right)^{-\nu}, \\
\tilde{D}_{\mu\mu}^{\text{2D}} &= 2C(\nu) \frac{1-\mu^2}{R^2} \frac{\delta B_{\text{2D}}^2}{B_0^2} \int_0^\infty \mathrm{d}x \; (1+x^2)^{-\nu} \\
&\quad \times \frac{\tilde{D}_\perp \xi^2 x^2}{\frac{1-\mu^2}{2}\xi^2 x^2 + \left[\tilde{D}_\perp \xi^2 x^2\right]^2 + R^{-2}}, \\
\tilde{D}_\perp^{\text{slab}} &= 0, \\
\tilde{D}_\perp^{\text{2D}} &= 2C(\nu)\mu^2 R \frac{\delta B_{\text{2D}}^2}{B_0^2} \int_0^\infty \mathrm{d}x \; (1+x^2)^{-\nu} \cdot \tilde{V} \cdot H(\tilde{V}, \tilde{W}),
\end{aligned} \tag{7.65}
$$

with

$$
\begin{aligned}
\tilde{V} &= R\tilde{D}_\perp \xi^2 x^2 + R\tilde{\omega}, \\
\tilde{W} &= R\xi x\sqrt{1-\mu^2}.
\end{aligned} \tag{7.66}
$$

Approximations for the function $H(\tilde{V}, \tilde{W})$ are given in (5.39). This modification (or extension) of the weakly nonlinear theory is important only for perpendicular transport in slab-like turbulence models. Therefore, the results of the standard WNLT should be a good approximation.

Chapter 8
Applications

In this chapter, three examples for the application of nonlinear diffusion theory are discussed. First, we investigate particle transport in the heliosphere by revisiting the Palmer consensus. It is demonstrated that the observed perpendicular mean free paths can be reproduced by the NLGC theory if the nonlinear anisotropic dynamical turbulence model is used. In the second part of this chapter, we employ the NLGC theory for describing particle acceleration at highly perpendicular interplanetary shocks. As a third example, the problem of the observed primary-to-secondary abundance ratio is revisited, and it is shown that the weakly nonlinear approach can reproduce these measurements.

8.1 Particle Transport in the Heliosphere

The understanding of the interaction between cosmic rays and turbulent fields of the solar wind plasma is one of the fundamental problems of space physics. An early treatment of particle transport employed the quasilinear theory for a simple magnetostatic slab model (standard QLT, see Chap. 3). Palmer (1982) compared the predictions of this model for the parallel mean free path with heliospheric observations (see Fig. 1.5) and noted two major problems:

1. *Magnitude problem:* Typically, the observed parallel mean free paths are much larger than the predicted standard QLT results.
2. *Flatness problem:* The observed parallel mean free paths are generally constant with a rigidity-independent mean free path for 0.5–5,000 MV, but standard QLT predicts that the mean free path should increase with increasing rigidity ($\lambda_{\parallel} \sim R^{1/3}$).

Bieber et al. (1994) used the slab/2D composite model to achieve agreement between theory and observations. Furthermore, Bieber et al. (1994) used the so-called damping model of dynamical turbulence (DT model, see Chap. 2), where the dynamical correlation function is an exponential function. Although the agreement between this model and observations is acceptable (see also Dröge 2000), there are several unanswered questions. First, the DT model applied in Bieber et al. (1994) and Dröge (2000) is a simple model for the dynamical correlation function and can

A. Shalchi, *Nonlinear Cosmic Ray Diffusion Theories*,
Astrophysics and Space Science Library,
DOI: 10.1007/978-3-642-00309-7_8,

be seen as a crude approximation. Furthermore, the DT model does not take into account the plasma wave character of the turbulence. The third problem is the most difficult problem: it was demonstrated in Chaps. 3–7 that nonlinear effects play a crucial role if diffusion coefficients are calculated. These effects have been neglected due to the application of QLT in Bieber et al. (1994) and Dröge (2000).

In this chapter, we revisit the Palmer consensus by applying the nonlinear anisotropic dynamical turbulence model (NADT model, see Chap. 2) for the dynamical correlation function. It is the purpose of this chapter to compute scattering mean free paths by using the NADT model and to compare theoretical results of the NADT model with other models and heliospheric observations. Besides the parallel mean free path, the mean free path perpendicular to the background magnetic field can also be obtained from observations. In this chapter, the NLGC approach is applied because this theory is more tractable than WNLT. For the turbulence geometry, we employ the slab/2D composite model and, for the wave spectrum, we use an improved form in comparison to the spectrum discussed in Chap. 2 with energy, inertial, and dissipation range:

$$g^{\text{slab}}(k_\parallel) = \frac{C(\nu)}{2\pi} l_{\text{slab}} \delta B_{\text{slab}}^2 \begin{cases} (1 + k_\parallel^2 l_{\text{slab}}^2)^{-\nu} & \text{for } k_\parallel \leq k_{\text{slab}} \\ (1 + k_{\text{slab}}^2 l_{\text{slab}}^2)^{-\nu} (k_{\text{slab}}/k_\parallel)^p & \text{for } k_\parallel \geq k_{\text{slab}}, \end{cases} \tag{8.1}$$

$$g^{2\text{D}}(k_\perp) = \frac{2C(\nu)}{\pi} l_{2\text{D}} \delta B_{2\text{D}}^2 \begin{cases} (1 + k_\perp^2 l_{2\text{D}}^2)^{-\nu} & \text{for } k_\perp \leq k_{2\text{D}} \\ (1 + k_{2\text{D}}^2 l_{2\text{D}}^2)^{-\nu} (k_{2\text{D}}/k_\perp)^p & \text{for } k_\perp \geq k_{2\text{D}}. \end{cases} \tag{8.2}$$

Such a spectrum is more in agreement with the observed spectra (see Fig. 2.3). The spectrum used is shown in Fig. 8.1 and was originally proposed by Bieber et al. (1994). Besides the turbulence parameters used in the standard spectrum (these parameters are explained in Chap. 2), we introduced the dissipation range spectral index p and the dissipation wavenumber k_i. For two-component turbulence and by using the NADT model, the turbulence correlation tensor that enters transport theories has the form

$$P_{lm}(\mathbf{k}, t) = P_{lm}^{\text{slab}}(\mathbf{k})\, \Gamma^{\text{slab}}(k_\parallel, t) + P_{lm}^{2\text{D}}(\mathbf{k})\, \Gamma^{2\text{D}}(k_\perp, t), \tag{8.3}$$

with $P_{lm}^{\text{slab}}(\mathbf{k})$ from (2.32), $P_{lm}^{2\text{D}}$ from (2.50), $\Gamma^{\text{slab}}(k_\parallel, t)$ from (2.123), and $\Gamma^{2\text{D}}(k_\perp, t)$ from (2.126). In the following, (8.3) is combined with QLT to compute parallel mean free paths and with NLGC theory to determine perpendicular diffusion coefficients.

8.1.1 The Quasilinear Parallel Mean Free Path

Here we employ QLT (see Chap. 3) to determine the parallel mean free path. Because of the dynamical effects, we have the following resonance function:

$$R_n^i(\mathbf{k}) = \Re\mathfrak{e} \int_0^\infty dt\, e^{-i(k_\parallel v_\parallel + n\Omega)t}\, \Gamma^i(\mathbf{k}, t). \tag{8.4}$$

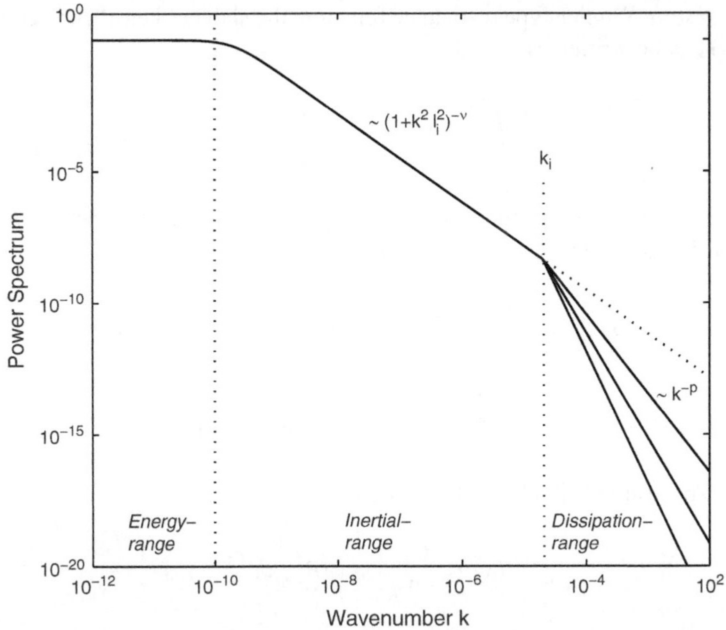

Fig. 8.1 The wave spectrum used for the description of heliospheric cosmic ray propagation. A general spectrum with energy, inertial, and dissipation range is used. The dissipation wavenumber k_i (with i = slab, 2D) divides the inertial range from the dissipation range

By applying (2.123) and (2.126) for the functions $\Gamma^i(\mathbf{k}, t)$, we obtain a Breit–Wigner-type resonance function within the NADT model and QLT. Similar forms have been derived in Chap. 5 by using WNLT. In this case, resonance broadening is caused by nonlinear effects (caused by pitch-angle and perpendicular diffusion), whereas, in the present calculations, resonance broadening is caused by the dynamical turbulence effects. However, the analytical derivation of pitch-angle Fokker–Planck coefficients presented in the following can be performed in the same manner as the weakly nonlinear calculations due to the similar structure of the resonance functions.

8.1.1.1 The Pitch-angle Fokker–Planck Coefficient for Pure Slab Geometry

The resonance function of (8.4) for pure slab geometry, with (2.123) for $\Gamma^{\text{slab}}(\mathbf{k}, t)$, is

$$\begin{aligned} R_n^{\text{slab}} &= \Re\mathfrak{e} \int_0^\infty \mathrm{d}t \; \mathrm{e}^{-\mathrm{i}(k_\parallel v_\parallel + n\Omega - v_A k_\parallel)t - \beta t}, \\ &= \frac{\beta}{\beta^2 + (k_\parallel v_\parallel + n\Omega - v_A k_\parallel)^2}. \end{aligned} \tag{8.5}$$

With this Breit–Wigner-type resonance function, the slab Fokker–Planck coefficient of (3.34) can be written as

$$D_{\mu\mu}^{\text{slab}} = \frac{2\pi\Omega^2(1-\mu^2)}{B_0^2}\int_0^\infty \mathrm{d}k_\parallel\, g^{\text{slab}}(k_\parallel)\sum_{n=\pm 1}\frac{\beta}{\beta^2+(k_\parallel v_\parallel + n\Omega - v_{\text{A}}k_\parallel)^2}. \tag{8.6}$$

With the integral transformation $x = l_{\text{slab}}k_\parallel$ and with the parameters $R = R_{\text{L}}/l_{\text{slab}} = v/(\Omega l_{\text{slab}})$ and $\varepsilon = v_{\text{A}}/v$, we obtain

$$D_{\mu\mu}^{\text{slab}} = \frac{2\pi(1-\mu^2)}{B_0^2 l_{\text{slab}}}\int_0^\infty \mathrm{d}x\, g^{\text{slab}}\left(k_\parallel = \frac{x}{l_{\text{slab}}}\right) \times \sum_{n=\pm 1}\frac{\beta}{(\beta/\Omega)^2+[xR(\mu-\varepsilon)+n]^2}. \tag{8.7}$$

The slab spectrum of (2.33) can be written as

$$g^{\text{slab}}(x) = \frac{C(\nu)}{2\pi} l_{\text{slab}}\delta B_{\text{slab}}^2\, h^{\text{slab}}(x), \tag{8.8}$$

with

$$h^{\text{slab}}(x) = \begin{cases} (1+x^2)^{-\nu} & \text{for } x \leq \xi_{\text{slab}} \\ (1+\xi_{\text{slab}}^2)^{-\nu}(\frac{\xi_{\text{slab}}}{x})^p & \text{for } x \geq \xi_{\text{slab}}, \end{cases} \tag{8.9}$$

where we used $\xi_{\text{slab}} = l_{\text{slab}}k_{\text{slab}}$. Then we find for the dimensionless Fokker–Planck coefficient

$$\begin{aligned} \tilde{D}_{\mu\mu}^{\text{slab}} &= \frac{l_{\text{slab}}}{v}D_{\mu\mu}^{\text{slab}} \\ &= \frac{C(\nu)(1-\mu^2)}{R}\frac{\delta B_{\text{slab}}^2}{B_0^2}\int_0^\infty \mathrm{d}x\, h^{\text{slab}}(x) \\ &\quad \times \sum_{n=\pm 1}\frac{\beta/\Omega}{(\beta/\Omega)^2+[xR(\mu-\varepsilon)+n]^2}. \end{aligned} \tag{8.10}$$

The parameter β/Ω can be expressed as

$$\beta/\Omega = \sqrt{2}\alpha\varepsilon\rho\frac{\delta B_{\text{2D}}}{B_0}R \equiv \sqrt{2}\alpha\frac{v_{\text{A}}}{v}\frac{R_{\text{L}}}{l_{\text{2D}}}\frac{\delta B_{\text{2D}}}{B_0}, \tag{8.11}$$

where we used $\rho = l_{\text{slab}}/l_{\text{2D}}$. By using the analytical approximations presented in Shalchi et al. (2006, Appendix), a numerical evaluation of (8.10) is a simple matter.

8.1.1.2 The Pitch-angle Fokker–Planck Coefficient for Pure 2D

For pure 2D turbulence, the pitch-angle Fokker–Planck coefficient has the form (see (3.53))

$$D_{\mu\mu}^{\mathrm{2D}} = \frac{2\pi\Omega^2(1-\mu^2)}{B_0^2}\int_0^\infty \mathrm{d}k_\perp\, g^{\mathrm{2D}}(k_\perp)\sum_{n=-\infty}^{\infty} R_n^{\mathrm{2D}}(k_\perp)\frac{n^2 J_n^2(W)}{W^2}, \tag{8.12}$$

with $W = k_\perp R_{\mathrm{L}}\sqrt{1-\mu^2}$. For pure 2D, the resonance function becomes

$$R_n^{\mathrm{2D}} = \Re\mathfrak{e}\int_0^\infty \mathrm{d}t\, \mathrm{e}^{-\mathrm{i}n\Omega t-\gamma^{\mathrm{2D}}t} = \frac{\gamma^{\mathrm{2D}}}{(\gamma^{\mathrm{2D}})^2+(n\Omega)^2}, \tag{8.13}$$

and we obtain for the pitch-angle Fokker–Planck coefficient

$$D_{\mu\mu}^{\mathrm{2D}} = \frac{2\pi\Omega^2(1-\mu^2)}{B_0^2}\int_0^\infty \mathrm{d}k_\perp\, g^{\mathrm{2D}}(k_\perp)\sum_{n=-\infty}^{\infty}\frac{\gamma^{\mathrm{2D}}}{(\gamma^{\mathrm{2D}})^2+(n\Omega)^2}\frac{n^2 J_n^2(W)}{W^2}. \tag{8.14}$$

With the approximation of (5.29), $D_{\mu\mu}^{\mathrm{2D}}$ can be simplified to

$$D_{\mu\mu}^{\mathrm{2D}} = \frac{2\pi(1-\mu^2)}{B_0^2}\int_0^\infty \mathrm{d}k_\perp\, g^{\mathrm{2D}}(k_\perp)\frac{\gamma^{\mathrm{2D}}}{W^2+2(\gamma^{\mathrm{2D}}/\Omega)^2+2}. \tag{8.15}$$

Using the integral transformation $x = k_\perp l_{\mathrm{2D}}$ and $\rho = l_{\mathrm{slab}}/l_{\mathrm{2D}}$, we deduce

$$\begin{aligned} D_{\mu\mu}^{\mathrm{2D}} =& \frac{2\pi(1-\mu^2)}{B_0^2 l_{\mathrm{2D}}}\int_0^\infty \mathrm{d}x\, g^{\mathrm{2D}}\left(k_\perp = \frac{x}{l_{\mathrm{2D}}}\right) \\ &\times \frac{\gamma^{\mathrm{2D}}}{\left[xR\rho\sqrt{1-\mu^2}\right]^2+2(\gamma^{\mathrm{2D}}/\Omega)^2+2}. \end{aligned} \tag{8.16}$$

The 2D spectrum of (2.51) can be written as

$$g^{\mathrm{2D}} = \frac{2C(\nu)}{\pi} l_{\mathrm{2D}}\delta B_{\mathrm{2D}}^2\, h^{\mathrm{2D}}(x), \tag{8.17}$$

with

$$h^{\mathrm{2D}}(x) = \begin{cases} (1+x^2)^{-\nu} & \text{for } x \le \xi_{\mathrm{2D}} \\ (1+\xi_{\mathrm{2D}}^2)^{-\nu}\left(\frac{\xi_{\mathrm{2D}}}{x}\right)^p & \text{for } x \ge \xi_{2D}, \end{cases} \tag{8.18}$$

where we used $\xi_{2D} = l_{2D} k_{2D}$. Then we find for the dimensionless Fokker–Planck coefficient

$$\tilde{D}_{\mu\mu}^{2D} = \frac{4C(\nu)(1-\mu^2)}{R} \frac{\delta B_{2D}^2}{B_0^2} \int_0^\infty dx\, h^{2D}(x) \times \frac{\tilde{\gamma}^{2D}}{\left[xR\rho\sqrt{1-\mu^2}\right]^2 + 2(\tilde{\gamma}^{2D})^2 + 2}. \tag{8.19}$$

The parameter $\tilde{\gamma}^{2D}$ can be written as

$$\tilde{\gamma}^{2D} = \frac{\gamma^{2D}}{\Omega} = \frac{\beta}{\Omega} \begin{cases} 1 & \text{for } x \leq 1 \\ x^{2/3} & \text{for } x \geq 1. \end{cases} \tag{8.20}$$

The integral in (8.19) can be solved numerically. Together with the slab results of the previous paragraph, we can determine the total Fokker–Planck coefficient in composite slab/2D turbulence. If we express the parallel mean free path by the dimensionless pitch-angle Fokker–Planck coefficients, we have for composite geometry

$$\lambda_\parallel = \frac{3}{4} l_{\text{slab}} \int_0^1 d\mu\, \frac{(1-\mu^2)^2}{\tilde{D}_{\mu\mu}^{\text{slab}}(\mu) + \tilde{D}_{\mu\mu}^{2D}(\mu)}. \tag{8.21}$$

To replace $\tilde{D}_{\mu\mu}^{\text{slab}}$ and $\tilde{D}_{\mu\mu}^{2D}$, we can use (8.10) and (8.19).

8.1.2 The Nonlinear Perpendicular Mean Free Path

So far, we have considered only parallel spatial diffusion, but particle propagation in the direction perpendicular to the background field is not less important. Here, we employ the NLGC theory (see Chap. 4) to compute the perpendicular mean free path. In NLGC theory, the parallel mean free path is simply an input parameter for an integral equation that determines the perpendicular mean free path. For dynamical turbulence, the NLGC theory has to be reformulated (see e.g., Shalchi et al. 2004c). Starting with (4.45), we have with (8.3)

$$\kappa_\perp = \frac{a^2 v^2}{3B_0^2} \int d^3k\, P_{xx}^{\text{slab}}(\mathbf{k}) \int_0^\infty dt\, e^{-vt/\lambda_\parallel - k_\perp^2 \kappa_\perp t - k_\parallel^2 \kappa_\parallel t} \Re\left[\Gamma^{\text{slab}}(\mathbf{k}, t)\right] + \frac{a^2 v^2}{3B_0^2} \int d^3k\, P_{xx}^{2D}(\mathbf{k}) \int_0^\infty dt\, e^{-vt/n\lambda_\parallel - k_\perp^2 \kappa_\perp t - k_\parallel^2 \kappa_\parallel t} \Re\left[\Gamma^{2D}(\mathbf{k}, t)\right], \tag{8.22}$$

where we assumed axisymmetric turbulence ($\kappa_\perp = \kappa_{xx} = \kappa_{yy}$). Applying the NADT model

$$\begin{aligned} \mathfrak{Re}\left[\Gamma^{\text{slab}}(\mathbf{k},t)\right] &= \cos(\omega t)\,\mathrm{e}^{-\gamma^{\text{slab}}t}, \\ \mathfrak{Re}\left[\Gamma^{\text{2D}}(\mathbf{k},t)\right] &= \mathrm{e}^{-\gamma^{\text{2D}}t} \end{aligned} \tag{8.23}$$

and, by using

$$\begin{aligned} &\int_0^\infty \mathrm{d}t\, \mathrm{e}^{-vt/\lambda_\parallel - k_\perp^2\kappa_\perp t - k_\parallel^2\kappa_\parallel t} \cos(\omega t)\,\mathrm{e}^{-\gamma t} \\ &\quad = \frac{v/\lambda_\parallel + k_\perp^2\kappa_\perp + k_\parallel^2\kappa_\parallel + \gamma}{(v/\lambda_\parallel + k_\perp^2\kappa_\perp + k_\parallel^2\kappa_\parallel + \gamma)^2 + \omega^2}, \end{aligned} \tag{8.24}$$

we obtain

$$\begin{aligned} \kappa_\perp = {}& \frac{a^2v^2}{3B_0^2}\int \mathrm{d}^3k\; P_{xx}^{\text{slab}}(\mathbf{k}) \frac{v/\lambda_\parallel + k_\perp^2\kappa_\perp + k_\parallel^2\kappa_\parallel + \gamma^{\text{slab}}}{(v/\lambda_\parallel + k_\perp^2\kappa_\perp + k_\parallel^2\kappa_\parallel + \gamma^{\text{slab}})^2 + \omega^2} \\ &+ \frac{a^2v^2}{3B_0^2}\int \mathrm{d}^3k\; P_{xx}^{\text{2D}}(\mathbf{k}) \frac{1}{v/\lambda_\parallel + k_\perp^2\kappa_\perp + k_\parallel^2\kappa_\parallel + \gamma^{\text{2D}}}. \end{aligned} \tag{8.25}$$

For P_{xx}^{slab} and P_{xx}^{2D}, we have (see (2.32) and (2.50))

$$\begin{aligned} P_{xx}^{\text{slab}}(\mathbf{k}) &= g^{\text{slab}}(k_\parallel)\frac{\delta(k_\perp)}{k_\perp} \\ P_{xx}^{\text{2D}}(\mathbf{k}) &= g^{\text{2D}}(k_\perp)\frac{\delta(k_\parallel)}{k_\perp}\,\sin^2\Psi, \end{aligned} \tag{8.26}$$

with $\sin\Psi = k_y/k_\perp$. With the two wave spectra from (2.33) and (2.51) and, by applying $\lambda_\parallel = 3\kappa_\parallel/nv$ and $\lambda_\perp = 3\kappa_\perp/v$, we finally find

$$\begin{aligned} \frac{\lambda_\perp}{\lambda_\parallel} = {}& 2a^2C(\nu)\frac{\delta B_{\text{slab}}^2}{B_0^2}\int_0^\infty \mathrm{d}x\; h^{\text{slab}}(x) \\ &\times \frac{1 + \lambda_\parallel^2/(3l_{\text{slab}}^2)\,x^2 + \lambda_\parallel/nv\;\gamma^{\text{slab}}}{(1 + \lambda_\parallel^2/(3l_{\text{slab}}^2)\,x^2 + \lambda_\parallel/v\;\gamma^{\text{slab}})^2 + (\varepsilon\lambda_\parallel/nl_{\text{slab}}\,x)^2} \\ &+ 2a^2C(\nu)\frac{\delta B_{\text{2D}}^2}{B_0^2}\int_0^\infty \mathrm{d}x\; h^{\text{2D}}(x)\frac{1}{1 + \lambda_\parallel\lambda_\perp/(3l_{\text{2D}}^2)\,x^2 + \lambda_\parallel/v\;\gamma^{\text{2D}}}. \end{aligned} \tag{8.27}$$

To replace γ^{slab} and γ^{2D}, we can use

$$\frac{\lambda_\parallel}{v}\gamma^{\text{slab}} = \frac{\lambda_\parallel}{Rl_{\text{slab}}}\frac{\beta}{\Omega} \tag{8.28}$$

and

$$\frac{\lambda_\parallel}{v}\gamma^{2\mathrm{D}} = \frac{\lambda_\parallel}{R l_{\mathrm{slab}}}\frac{\beta}{\Omega}\begin{cases}1 & \text{for } x \leq 1\\ x^{2/3} & \text{for } x \geq 1.\end{cases} \tag{8.29}$$

The parameter β is defined in (2.129) and for a^2 we assume $a^2 = 1/3$ as originally proposed in Matthaeus et al. (2003). A numerical investigation of (8.27) is straightforward, and the results will be presented together with the QLT results for parallel diffusion in the next section.

8.1.3 Numerical Results Obtained by Using the NADT Model

Here, we evaluate the formulas for the mean free paths derived in the previous paragraphs numerically for the parameter set of Table 8.1, which should be appropriate for interplanetary conditions at 1 AU heliocentric distance. All formulas depend on the parameter ε, which can be expressed as

$$\varepsilon = \frac{v_{\mathrm{A}}}{v} = \frac{v_{\mathrm{A}}}{c}\frac{\sqrt{R_0^2 + R^2}}{R}, \tag{8.30}$$

with

$$R_0 = \frac{1}{l_{\mathrm{slab}}\, B_0}\begin{cases}0.511\,\mathrm{MV} & \text{for electrons}\\ 938\,\mathrm{MV} & \text{for protons.}\end{cases} \tag{8.31}$$

Table 8.1 Parameters used for our numerical calculations

Parameter	Symbol	Value
Inertial range spectral index	2ν	$5/3$
Dissipation range spectral index	p	3
Alfvén speed	v_{A}	$33.5\,\mathrm{km\,s^{-1}}$
Slab bendover scale	l_{slab}	0.030 AU
2D bendover scale	$l_{2\mathrm{D}}$	$0.1\, l_{\mathrm{slab}}$
Slab dissipation wavenumber	k_{slab}	$3 \times 10^6\,(\mathrm{AU})^{-1}$
2D dissipation wavenumber	$k_{2\mathrm{D}}$	$3 \times 10^6\,(\mathrm{AU})^{-1}$
Mean field	B_0	4.12 nT
Turbulence strength	$\delta B / B_0$	1
Slab fraction	$\delta B^2_{\mathrm{slab}}$	$0.2\,\delta B^2$
2D fraction	$\delta B^2_{2\mathrm{D}}$	$0.8\,\delta B^2$

These values should be appropriate for the solar wind at 1 AU heliocentric distance. If a parameter is different from these values, we note this separately in the corresponding figures and discussions

For the heliospheric parameters considered here, we have for electrons $R_0(e^-) \approx 9.2 \times 10^{-5}$ and for protons $R_0(p^+) \approx 0.169$. In the plasma wave model, we must take into account the restriction $\varepsilon = v_A/v \ll 1$ (see e.g., Schlickeiser 2002). Thus, we can consider only rigidities that satisfy the following condition:

$$R \gg \frac{R_0}{\sqrt{(c/v_A)^2 - 1}} \approx R_0 \frac{v_A}{c}. \tag{8.32}$$

For $v_A = 33.5\,\mathrm{km\,s^{-1}}$, this becomes for electrons $R(\mathrm{e}^-) \gg 10^{-8}$ and for protons $R(p^+) \gg 2 \times 10^{-5}$. Therefore, the proton results presented in this book could be invalid for $R(p^+) \leq 2 \times 10^{-5}$.

In the following, we compare NADT results with observations and with other turbulence models. Furthermore, we discuss the dependence of $\lambda_\parallel$ and $\lambda_\perp$ upon different turbulence parameters.

8.1.3.1 Pitch-angle Diffusion Within the NADT Model

Figure 8.2 shows the pitch-angle Fokker–Planck coefficients calculated within the NADT model. In general, the pitch-angle Fokker–Planck coefficient is no longer equal to zero at 90° ($\mu = 0$), as was the case in the magnetostatic model, so that we no longer obtain an infinitely large parallel mean free path as in the magnetostatic case. It should be noted, however, that QLT itself is questionable close to 90°. By considering Fig. 8.2, we find that, at least for protons, pitch-angle scattering close to 90° is very strong due to the dynamical effects. Therefore, one could assume that nonlinear effects that also lead to nonvanishing pitch-angle scattering at 90° could be suppressed.

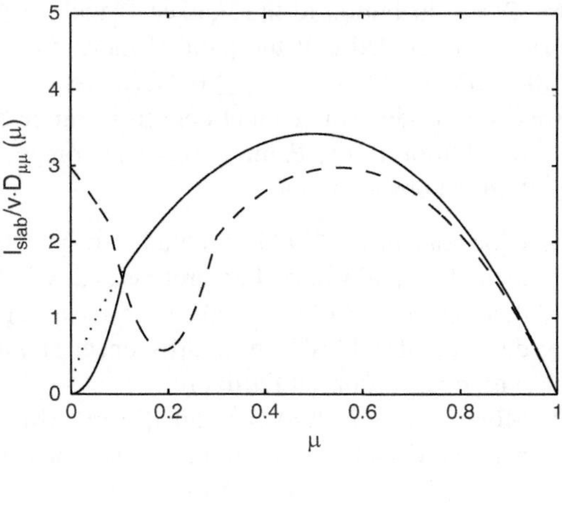

Fig. 8.2 The pitch-angle Fokker–Planck coefficient calculated by employing the NADT model for protons (*dashed line*) and electrons (*solid line*). Also the magnetostatic model is shown (*dotted line*). The panel shows the dimensionless pitch-angle Fokker–Planck coefficient $\widetilde{D}_{\mu\mu}$ as a function of the pitch-angle cosine μ. All results are for nearly pure slab geometry ($\delta B_{\mathrm{slab}}^2/B_0^2 = 0.9$) and for small rigidities $R = 10^{-4}$

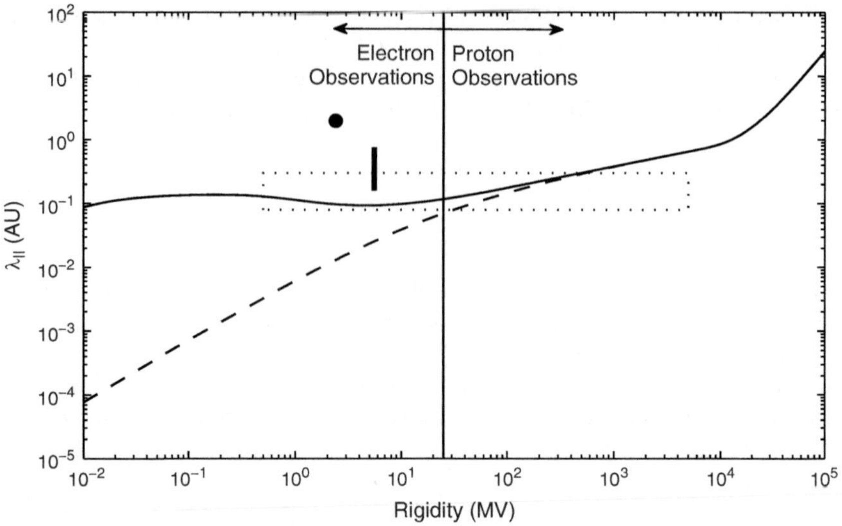

Fig. 8.3 The parallel mean free path $\lambda_{\|}$. Shown are QLT results for electrons (*solid line*) and protons (*dashed line*) in comparison with different observations: Palmer consensus (Palmer 1982, *box*), Ulysses observations (Gloeckler et al. 1995, *dot*), and AMPTE spacecraft observations (Möbius et al. 1998, *vertical line*)

8.1.3.2 Parallel Mean Free Paths in Comparison with the Palmer Consensus and Pickup Ion Observations

In Fig. 8.3, the parallel mean free path is shown in comparison with observations. Shown is the mean free path in astronomical units vs. the magnetic rigidity in MV units. The NADT results are not much different from results obtained by using the DT model (damping model of dynamical turbulence, Bieber et al. 1994), but here, the 2D contribution to the Fokker–Planck coefficient plays a crucial role. Palmer (1982) concluded that the parallel mean free path for rigidities between 0.5 and 5,000 MV is $0.08\,\text{AU} \leq \lambda_{\|} \leq 0.3\,\text{AU}$. It is one of the key results of this chapter that we can reproduce the Palmer consensus range by applying the NADT model.

In addition to the Palmer consensus range, our results are also compared with pickup ion observations:

1. Gloeckler et al. (1995) concluded from Ulysses observations that the parallel mean free paths of pickup protons is 2 AU at 2.4 MV rigidity. It should be noted that this observation was taken at high heliographic latitudes and at a heliocentric distance of 2.34 AU; these differences should be remembered when comparing with observations at Earth orbit.
2. Möbius et al. (1998) concluded from AMPTE spacecraft observations that the parallel mean free paths of pickup helium ranges from 0.16 to 0.76 AU at 5.6 MV rigidity in the data they analyzed in their article.

Both results are also shown in Fig. 8.3. The Möbius et al. (1998) observations are close to the theoretical results, but the Gloeckler et al. (1995) measurements are much larger than our theoretical predictions. This discrepancy might be due to the fact that the Gloeckler et al. (1995) observation was taken at high heliographic latitudes and at a heliocentric distance of 2.34 AU. For such conditions, the turbulence parameters are expected to be different from the values used for the theoretical calculations (see Table 8.1). For example, if we decrease the strength of turbulence $\delta B/B_0$, we would obtain a much larger parallel mean free path. Therefore, we expect that the Gloeckler et al. (1995) and the Möbius et al. (1998) observations can be reproduced by changing the turbulence parameters.

8.1.3.3 Perpendicular Mean Free Paths in Comparison with the Palmer Consensus, Jovian Electrons, and Galactic Protons

In Fig. 8.4, the perpendicular mean free paths at 1 AU, calculated with NLGC theory by using the NADT model, are compared with observations. Our new results agree very well with observational determinations from Jovian electrons (Chenette et al. 1977) and Ulysses measurements of galactic protons (Burger et al. 2000). Palmer (1982) concluded an average perpendicular diffusion coefficient at 1 AU of $\kappa_\perp\, c/v \approx 10^{21}\,\mathrm{cm}^2\,\mathrm{s}^{-1}$ and, therefore, $\lambda_\perp \approx 0.0067$ AU. In his article, the author pointed out that the spread around this average was rather large. By combining the average value for $\lambda_\perp$ with the Palmer consensus values for $\lambda_\parallel$, we find 0.02 AU $\leq \lambda_\perp \leq$ 0.083 AU. As illustrated by Fig. 8.4, the theoretical results deduced in this chapter agree very well with the Palmer consensus range. In Fig. 8.5, the ratio $\lambda_\perp/\lambda_\parallel$ is compared with the observations of Palmer (1982). By combining the average value for $\lambda_\perp$ with the Palmer consensus values for $\lambda_\parallel$, we find 0.02 AU $\leq \lambda_\perp/\lambda_\parallel \leq$ 0.083 AU. As illustrated by Fig. 8.5, the theoretical results agree very well with the Palmer consensus range. According to Fig. 8.5, the ratio of the perpendicular mean free path and the parallel mean free path satisfies the constraint

$$0.01 \leq \frac{\lambda_\perp}{\lambda_\parallel} \leq 0.1 \tag{8.33}$$

in the interesting energy regime where $R_\mathrm{L} \leq l_\mathrm{slab}$. For higher particle energies, we find even smaller values of this ratio.

8.1.3.4 The Influence of the Different Turbulence Parameters

A more detailed parameter study of NADT results is presented in Shalchi et al. (2006). The most important results are the following:

- Plasma wave effects described by the oscillation factor in Γ^{slab} can be neglected for nearly all parameter regimes

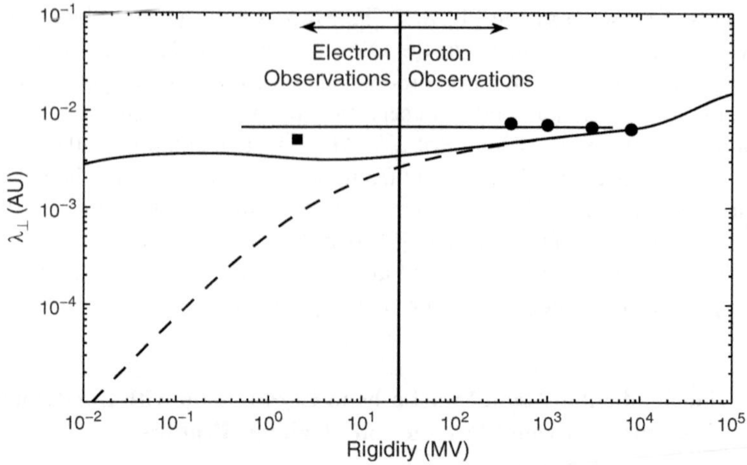

Fig. 8.4 The perpendicular mean free path $\lambda_\perp$. Shown are results for electrons (*solid line*) and protons (*dashed line*) in comparison with Jovian electrons (Chenette et al. 1977, *square*), Ulysses measurements of galactic protons (Burger et al. 2000, *dots*), and the Palmer (1982, *horizontal line*) value

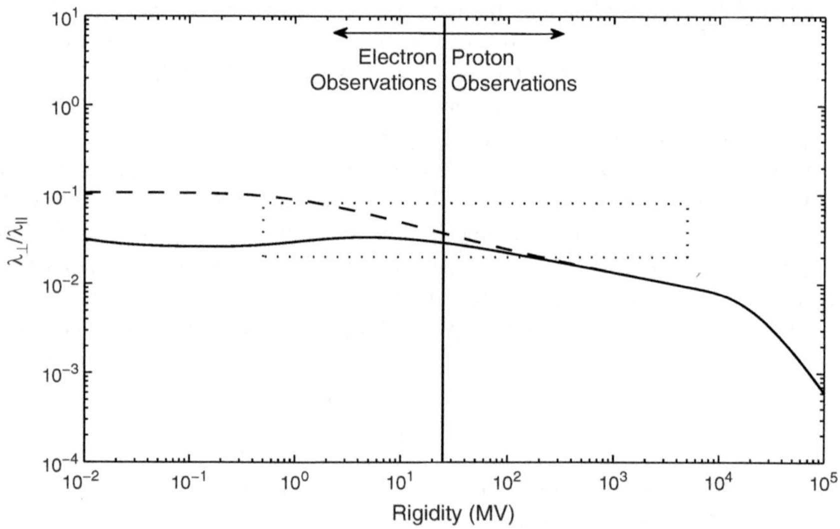

Fig. 8.5 The ratio $\lambda_\perp/\lambda_\parallel$. Shown are QLT results for electrons (*solid line*) and protons (*dashed line*) in comparison with observations (Palmer 1982, *box*)

- The ratio $l_{\mathrm{slab}}/l_{\mathrm{2D}}$ has a strong influence if the parallel mean free path is calculated by using the NADT model. The importance of this ratio indicates that the 2D contribution to the pitch-angle Fokker–Planck coefficient is important and must not be neglected even within QLT

- The results obtained by employing the DT model (a numerical evaluation within the DT model is presented in Bieber et al. (1994), and analytical formulas have been derived in Teufel and Schlickeiser (2002, 2003) and Shalchi and Schlickeiser (2004a)) provides results that are similar but not equal in comparison to the NADT model

8.1.4 Can We Indeed Reproduce Heliospheric Observations?

In this section, we employed the quasilinear approximation and the nonlinear guiding center theory to calculate the mean free paths by using the NADT model for the dynamical correlation function. A comparison with observations was also presented. We come to the following conclusions:

- Observed diffusion coefficients in the heliosphere can be explained by the NADT model.
- Within the NADT model, we can no longer neglect the 2D pitch-angle Fokker–Planck coefficient. Both the slab and 2D contributions to pitch-angle scattering are important.
- A nonlinear description of perpendicular transport is necessary. In this book, it is argued that QLT is not valid for transport perpendicular to the background magnetic field. For parallel transport, nonlinear effects can also be important, depending on the turbulence parameters.
- If the NADT model is an accurate turbulence model, we can neglect the contribution of the plasma wave dispersion relation to particle diffusion for most parameter regimes. Such effects are small in comparison to dynamical turbulence effects.
- The DT model that was considered in Bieber et al. (1994) and Dröge (2000) provides similar results for certain parameter regimes, but disagrees, in general, with the NADT model.
- It is demonstrated that the ratio of the bendover scales $l_{\mathrm{slab}}/l_{\mathrm{2D}}$ has a strong influence on the parallel mean free path. Also, the importance of nonlinear effects strongly depends on this ratio. An accurate determination of this ratio by observations would be an important step in space physics.

By combining the NADT model with QLT and the NLGC theory, we are, in principle, able to explain observed diffusion coefficients in the heliosphere. On the other hand, we can reproduce numerical test-particle simulations only by employing recently proposed nonlinear theories for parallel transport. It must be the subject of future work to find out whether nonlinear effects are important for parallel diffusion in dynamical turbulence.

8.2 Particle Acceleration at Perpendicular Shock Waves

Particle acceleration at shock waves occurs on scales ranging from cometary bow shocks (see e.g., Bavassano Cattaneo and Formisano 1994) to interplanetary shocks (see e.g., Zank et al. 2006) and supernova remnants (see e.g., Berezhko et al. 2002). The mechanism that is thought to be responsible for the observed power-law spectra of charged cosmic particles is diffusive shock acceleration (see e.g., Axford et al. 1977; Bell 1978a, b; Blandford and Ostriker 1978; Krymsky 1977).

The observed cosmic ray energy spectrum up to the knee can be explained by the particle acceleration at shock waves emitted by supernova remnants (see e.g., Völk et al. 1988). Furthermore, diffusive shock acceleration also accounts for gradual SEP (solar energetic particle) events (see e.g., Reames 1999), but the correspondence between the simple predictions of theory and observations has often not been compelling (see e.g., Desai et al. 2003) and more elaborate time-dependent models had to be developed (see e.g., Zank et al. 2000; Li et al. 2003, 2005; Rice et al. 2003; Li and Zank 2005).

Interplanetary shocks within 1 AU have regions that can be highly oblique or quasiparallel. Figure 8.6 shows a shock driven by a coronal mass ejection (CME) as an example. Clearly we can see the parallel as well as the perpendicular shock regimes. The importance of particle acceleration at perpendicular shocks has been recognized in the context of understanding both SEP events (shocks near the Sun) and energetic storm particle (ESP) events (shocks at 1 AU).

The theory of particle acceleration at a quasiparallel shock appears to be reasonably well understood, and has been applied to SEP and ESP events (see e.g. Zank et al. 2000; Li et al. 2003, 2005). More recently, Zank et al. (2006) developed and

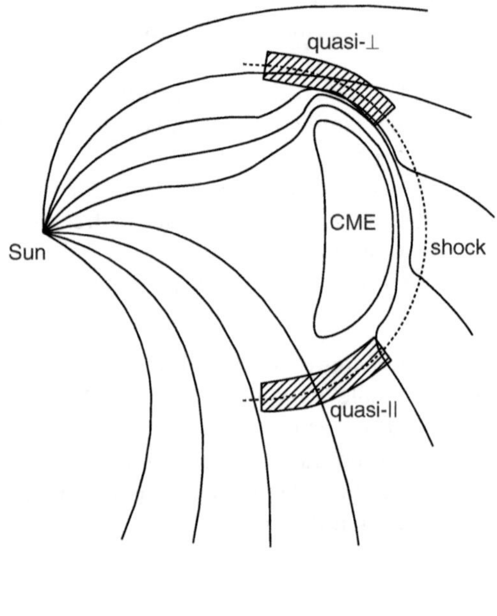

Fig. 8.6 Schematic of a CME (coronal mass ejection)-driven shock illustrating the variation in shock obliquity. Reprinted with permission from The American Geophysical Union – Zank et al. (2006)

described an approach for diffusive shock acceleration at perpendicular shocks. This approach is based on the NLGC theory discussed in Chap. 4. Some aspects of this approach will be discussed in the following paragraphs.

8.2.1 Interplanetary Shock Waves

Near an oblique shock front, Alfvén waves are responsible for particle scattering. Locally, at the simplest level, the particle distribution, f, and forward and backward wave energy densities, $I_\pm$, normalized to $B^2/(8\pi)$, per logarithmic bandwidth are coupled to each other through the one-dimensional transport equations for the particles and waves (Lee 1983; Bell 1978a, b; Gordon et al. 1999),

$$\frac{\partial f}{\partial t} = \frac{\partial}{\partial x}\left(\kappa_\parallel(p)\frac{\partial f}{\partial x}\right) - u\frac{\partial f}{\partial x} + \frac{p}{3}\frac{\partial u}{\partial x}\frac{\partial f}{\partial p}, \tag{8.34}$$

$$\frac{\partial I_\pm}{\partial t}(k,t) + (u \pm v_{\mathrm{A}})\frac{\partial I_\pm}{\partial x}(k,t) = \Gamma_\pm I_\pm(k,t) - \gamma_\pm I_\pm(k,t), \tag{8.35}$$

with the parallel spatial diffusion coefficient $\kappa_\parallel(p)$, the growth rates $\Gamma_\pm$ and $\gamma_\pm$ of $I_\pm$, the Alfvén speed v_{A}, and the flow speed u. In (8.34), the three terms on the right-hand side describe diffusion, convection due to the flow speed u, and adiabatic expansion/compression, respectively. Furthermore, the wavenumber $k = R_{\mathrm{L}}^{-1}$ with the particle Larmor radius R_{L} is used in (8.35). Equations (8.34) and (8.35) are appropriate to oblique shocks that are not very close to perpendicular. In this case, the particle transport equation and the wave energy equation are coupled through the spatial diffusion coefficient that depends on $I_\pm$, and these equations must be solved simultaneously to determine both the accelerated particle spectrum (responsible for generating waves) and the wave energy density (responsible for scattering the particles) self-consistently.

However, particle acceleration via the first-order Fermi mechanism at a perpendicular shock wave remains an outstanding problem for two essential reasons:

1. Unlike quasiparallel shocks, accelerated particles at a perpendicular shock cannot excite the (Alfvén) wave field that is responsible for scattering the particles repeatedly across the shock. This is because the growth term in the wave equation, besides being proportional to the particle gradient, is proportional to $\cos(\Theta_{\mathrm{bn}})$, where Θ_{bn} is the angle between the upstream magnetic field and the shock normal direction, and is thus zero for a strictly perpendicular shock (see e.g., Gordon et al., 1999; Li et al. 2003; Rice et al. 2003). Unlike the particle acceleration model described by (8.34) and (8.35), this therefore requires that particle scattering at a perpendicular shock be the result of in situ upstream turbulence that is convected into the shock. At the heart of the problem for perpendicular shocks is the need for a viable model of the perpendicular component of the diffusion tensor, which is discussed in greater detail in the next paragraph.

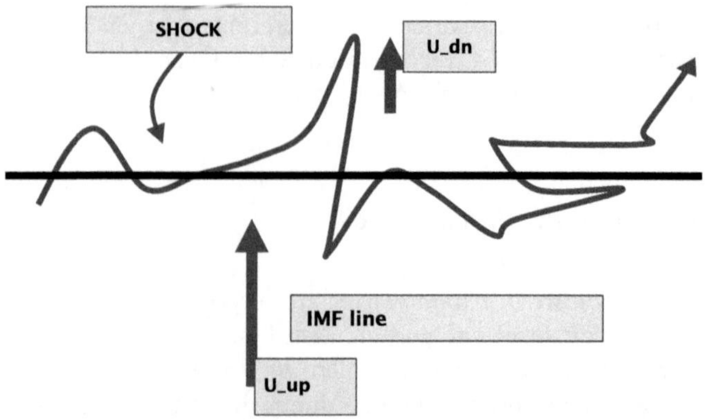

Fig. 8.7 Schematic showing an interplanetary magnetic field line experiencing random walking as it is convected through a shock. A particle attached to the field line can make multiple excursions upstream and down-stream as the field line is carried completely through the shock, provided the particle is moving sufficiently fast. This requires that particles already be energetic if they are to experience diffusive shock acceleration at a perpendicular shock. U_{up} and U_{dn} refer to the upstream and downstream flow speeds. Reprinted with permission from The American Geophysical Union – Zank et al. (2006)

2. Because an accelerated particle is essentially tied to a magnetic field line, its ability to cross the shock repeatedly is limited to the time it takes a magnetic field line to cross from upstream to downstream, assuming that the magnetic field line experiences some wandering to make the transmission time zero. Thus a fast-moving particle is necessary if it is to experience multiple crossings of the perpendicular shock (see Fig. 8.7) so that it can be diffusively accelerated. Consequently, diffusive shock acceleration is most effective for particles that are already energetic. This is referred to as the *injection problem* (see e.g., Jokipii 1987; Zank et al. 1996).

It is generally thought that the acceleration time at a perpendicular shock is much faster than at a parallel shock as the acceleration time scale (see e.g., Webb et al. 1995),

$$\tau_{\mathrm{acc}} = \left(\frac{1}{p}\frac{\mathrm{d}p}{\mathrm{d}t}\right)^{-1} = \frac{3s\kappa_{\perp}}{V_{\mathrm{sh}}^2(s-1)} \tag{8.36}$$

(where p is particle momentum, $\kappa_{\perp}$ is the spatial diffusion coefficient normal to the shock, s is the compression ratio, and V_{sh} is the upstream shock flow speed) is proportional to the particle spatial diffusion coefficient. In the following two paragraphs, we will restrict our investigation onto the computation of the acceleration time scale τ_{acc}. A more detailed investigation of particle acceleration at perpendicular shock waves and the calculation of the maximum particle energy can be found in Zank et al. (2006).

8.2.2 The Perpendicular Diffusion Coefficient

To describe particle acceleration at highly perpendicular shocks, we have to know the perpendicular diffusion coefficient (see (8.36)). In previous articles about shock acceleration, the so-called *Bohm limit* was employed. This limit can be obtained by assuming strong turbulence and therewith isotropic scattering of particles, that is, $\kappa_\parallel \approx \kappa_\perp$. Furthermore, it is assumed that the parallel mean free path $\lambda_\parallel = 3\kappa_\parallel/v$ cannot become shorter than the particle Larmor radius $\lambda_\parallel \geq R_{\rm L}$, and the perpendicular mean free path $\lambda_\perp = 3\kappa_\perp/v$ cannot become longer than the Larmor radius $\lambda_\perp \leq R_{\rm L}$. By additionally assuming that for strong turbulence the parallel mean free path becomes as short as possible and the perpendicular mean free path as long as possible, we find

$$\lambda^{\rm Bohm} := \lambda_\perp^{\rm Bohm} = \lambda_\parallel^{\rm Bohm} = R_{\rm L} \tag{8.37}$$

or, in terms of the diffusion coefficient,

$$\kappa^{\rm Bohm} = \frac{v}{3}\lambda^{\rm Bohm} = \frac{v}{3}R_{\rm L} = \frac{v^2}{3\Omega}. \tag{8.38}$$

By employing (1.8) for the unperturbed gyrofrequency Ω, one can easily derive

$$\kappa^{\rm Bohm} = \frac{v^2}{3}\frac{m\gamma c}{|q|\,B_0} = \frac{vpc}{3\,|q|\,B_0}. \tag{8.39}$$

Equation (8.39) is the standard expression for replacing the diffusion coefficient in the cosmic ray transport equation for investigating shock acceleration (see e.g., Berezhko and Völk 2007).

For describing cosmic ray perpendicular diffusion in the solar system, the FLRW limit derived in Chap. 3 (see (3.48))

$$\lambda_\perp^{\rm FLRW} = \frac{3\pi}{2}C(v)l_{\rm slab}\frac{\delta B_{\rm slab}^2}{B_0^2} \tag{8.40}$$

has been considered as the standard expression for a long time.

However, in Chaps. 4 and 7, the NLGC theory and its extended version for perpendicular scattering are derived, which are clearly in disagreement with the FLRW limit and the Bohm limit. According to Chap. 7 (see (7.54)), the perpendicular mean free path has the form[1]

---

[1] Here we employ the magnetostatic ENLGC-theory discussed in Chap. 7 to warrant a simple analytical estimation of the perpendicular diffusion coefficient and the acceleration time scale.

$$\lambda_{\perp}^{\mathrm{ENLGC}} = \left(\sqrt{3}\pi C(\nu) l_{2\mathrm{D}} \frac{\delta B_{2\mathrm{D}}^2}{B_0^2} \right)^{2/3} \lambda_{\parallel}^{1/3} \tag{8.41}$$

if $\lambda_{\parallel}\lambda_{\perp} \gg 3l_{2\mathrm{D}}^2$. To replace the parallel mean free path $\lambda_{\parallel}$, we employ the quasilinear results of Chap. 3 (see (3.43)). These results were derived by combining the quasilinear approximation with a magnetostatic model. For higher particle energies, this approximation should be valid (see discussions in Chaps. 3–6). Formula (3.43) provides a simple analytical expression for the parallel mean free path for particle energies for which dynamical turbulence effects as well as nonlinear effects are suppressed. We expect to find this regime for higher particle energies where the Larmor radius is larger than the slab bendover scale of the turbulence. A similar formula has been used in Zank et al. (2006, (10)). Together with (8.41), we can achieve a full analytical rigidity dependent description of perpendicular scattering. By combining (8.41) and (3.43), we find

$$\begin{aligned} \lambda_{\perp}^{\mathrm{QLT/ENLGC}} &= \left(\sqrt{3}\pi C(\nu) l_{2\mathrm{D}} \frac{\delta B_{2\mathrm{D}}^2}{B_0^2} \right)^{2/3} \\ &\times \left\{ \frac{3 l_{\mathrm{slab}}}{16\pi C(\nu)} \left(\frac{B_0}{\delta B_{\mathrm{slab}}} \right)^2 R^{2-2\nu} \left[\frac{2}{(1-\nu)(2-\nu)} + R^{2\nu} \right] \right\}^{1/3} . \end{aligned} \tag{8.42}$$

In Fig. 8.8, we compare the following analytical results with each other: the parallel mean free path from (3.43), the Bohm limit from (8.37), the FLRW limit from (8.40), and the combined QLT/ENLGC theory result from (8.42). Shown are the mean free paths normalized to the slab bendover scale l_{slab} vs. the ratio $R_{\mathrm{L}}/l_{\mathrm{slab}}$. In all cases considered, we have $\lambda_{\parallel}\lambda_{\perp} \gg 3l_{2\mathrm{D}}^2$ and, therefore, (8.41) and (8.42) are valid. For the parameter values, we have used the same values as in the other chapters of this book for interplanetary conditions (see e.g., Table 8.1). Clearly the nonlinear perpendicular mean free path is about a factor 10 shorter than the Bohm limit at high particle energies. In comparison to the FLRW limit, the nonlinear perpendicular mean free path is much longer. For low energetic particles, the Bohm limit provides a result that is too small. According to Fig. 8.8, the Bohm limit is not appropriate for replacing the perpendicular diffusion coefficient in the transport equation. In the context of acceleration at highly perpendicular interplanetary shocks, the idea of replacing the Bohm limit by a result obtained from a nonlinear diffusion theory was used by Zank et al. (2006).

8.2.3 *The Shock Acceleration Time Scale*

To evaluate the upper limit for the diffusively accelerated spectrum, we need to compute the shock acceleration time scale by employing (8.36). Theshock acceleration

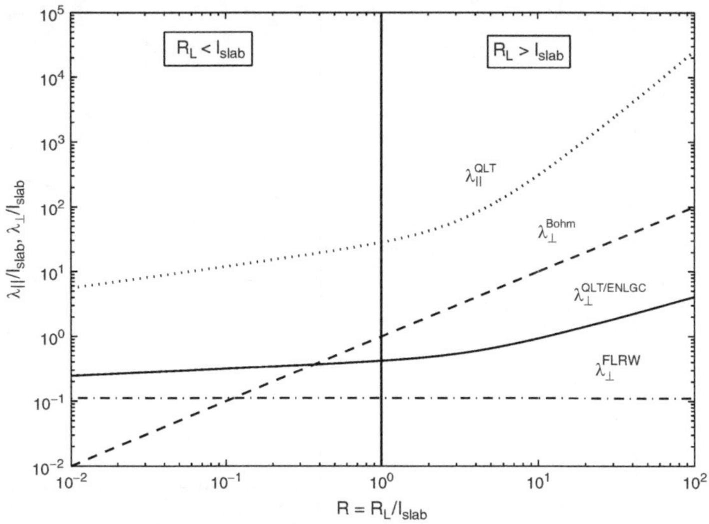

Fig. 8.8 Different results for the perpendicular mean free path: the Bohm limit (*dashed-line*), the FLRW limit (*dash-dotted line*), and the combined QLT/ENLGC theory result (*solid line*). Also shown is the QLT result for the parallel mean free path (*dotted line*)

time scale can easily be expressed through the perpendicular mean free path

$$\tau_{\rm acc} = \tau_0 \frac{R}{\sqrt{R_0^2 + R^2}} \frac{\lambda_\perp(R)}{l_{\rm slab}}, \tag{8.43}$$

where we have used

$$\frac{v}{c} = \frac{R}{\sqrt{R_0^2 + R^2}} \tag{8.44}$$

and

$$\tau_0 = \frac{c s l_{\rm slab}}{V_{\rm sh}^2 (s-1)}. \tag{8.45}$$

In (8.43) and (8.44), we used the parameter R_0 defined in (8.31). In Fig. 8.9, we have calculated the normalized shock acceleration time scale $\tau_{\rm acc}/\tau_0$ vs. the parameter $R = R_{\rm L}/l_{\rm slab}$ for protons (where we have $R_0 \approx 0.169$). Clearly we can see that the acceleration time scale obtained by employing (8.42) for $\lambda_\perp$ at high energies is significantly shorter than the Bohm acceleration time scale, which implies rapid acceleration at perpendicular shocks.

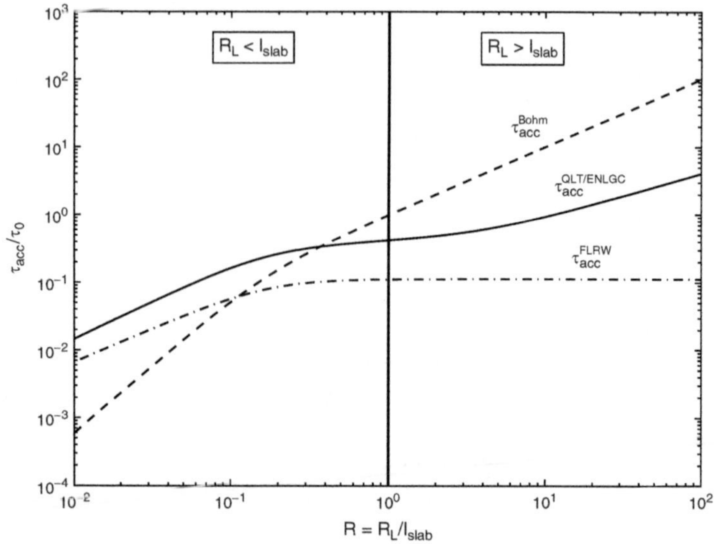

Fig. 8.9 Different results for the shock acceleration time scale normalized to τ_0: the Bohm limit (*dashed-line*), the FLRW limit (*dash-dotted line*), and the combined QLT/ENLGC theory result (*solid line*)

8.2.4 *Influence of Nonlinear Diffusion on Shock Acceleration*

In this chapter, we employed the nonlinear diffusion theory for describing particle acceleration at highly perpendicular shocks following the analysis of Zank et al. (2006). For a long time, particle acceleration at a perpendicular shock has been an outstanding problem, in large part due to the poor understanding of the perpendicular diffusion coefficient. By employing a linear description of parallel scattering (which can be justified by focusing on high-energetic particles) and the extended nonlinear guiding center theory (which provides a simple analytical expression for the parameter $\lambda_\perp$), we computed the particle acceleration time scale. This result, which is very similar to the result obtained by Zank et al. (2006), is also compared with the time scales obtained by employing the FLRW limit and the Bohm limit that has been used in previous articles about particle acceleration at shock waves. As shown in Fig. 8.9, the shock acceleration time scale at high energies is significantly smaller than the Bohm acceleration time scale, which implies rapid acceleration at perpendicular shocks.

However, the influence of realistic forms of the diffusion coefficient on particle acceleration at parallel shocks remains unclear. According to Chap. 5, nonlinear effects are suppressed for very high-energetic particles, and, therefore, QLT should be accurate. However, this statement is only correct for pure slab and slab/2D composite turbulence. For other turbulence geometries, 90° scattering is important (see Chap. 6). It will be the subject of future work to compute nonlinearly the parallel mean free path to achieve a better description of particle acceleration at parallel shocks. Also the influence of the rigidity dependence of the mean free paths onto particle acceleration processes in supernova shock waves is unclear and has to be explored in the future.

8.3 Primary-to-Secondary Abundance Ratio of Galactic Cosmic Rays

In Sect. 1.7, it has been described how the parallel mean free path controls the primary-to-secondary abundance ratio of galactic cosmic rays. According to observations (see e.g., Swordy et al. 1990), we have

$$\lambda_{\parallel}^{\text{obs}} \sim R^a, \quad a = 0.6 \pm 0.1. \tag{8.46}$$

In plain diffusion or leaky-box transport models without distributed stochastic acceleration, the implied value of $a = 0.6$ is not in accordance with the prediction $a = 1/3$ of the quasilinear theory with a Kolmogorov power spectrum of magnetic fluctuations $s = 5/3$ (see e.g., Chap. 3). For this reason, models with distributed stochastic acceleration have been favored (e.g., Jones et al. 2001) despite the disagreement of the implied weak energy dependence of the primary-to-secondary ratios at energies $\geq$20 GeV/nucleon with the high-energy HEAO-3 data by Binns et al. (1981) on the sub-Fe/Fe ratio.

In standard diffusion models, however, QLT is not appropriate to reproduce the observed abundance ratio of primary-to-secondary cosmic rays. But, as shown in Chap. 5, the weakly nonlinear theory provides a much steeper rigidity dependence. WNLT results depend on different turbulence parameters such as $\delta B/B_0$ and $\delta B_{\text{slab}}/\delta B$. Thus, we have to discuss the turbulence parameters of the interstellar medium, that control the mean free paths of galactic cosmic rays, before we compare WNLT with (8.46).

Estimates of interstellar magnetic field fluctuations (Minter and Spangler 1996; Beck et al. 1999) from Faraday rotation measure fluctuations, and radio polarization studies indicate rather large values between $0.1 \leq (\delta B/B_0)^2 < 0.65$. For such large turbulence values, nonlinear transport effects become important for calculating cosmic ray diffusion coefficients. For the other turbulence parameters, we use the values listed in Table 8.2.

Some of these parameters cannot be obtained from observations. Here, we simply adopt the values of the interplanetary medium. For instance, we have, according to Bieber et al. (1994, 1996), a slab/2D ratio $\delta B_{\text{slab}}^2/\delta B_{\text{2D}}^2 \approx 0.25$. For the spectral index of the inertial range, we assume $s = 2\nu = 5/3$ in accord with the Kolmogorov (1941) turbulence theory and heliospheric observations (see Chap. 2). The value $l_{\text{2D}} \approx 0.1\ l_{\text{slab}}$ has been found in laboratory experiments in an evolving magnetofluid plasma (Robinson and Rusbridge 1971). By using these values for the turbulence parameters in combination with WNLT, we calculate galactic

Table 8.2 Shown are the turbulence parameters used for the calculations presented in this section

Parameter	Symbol	Value
Intertial range spectral index	2ν	$5/3$
Slab bendover scale	l_{slab}	2×10^{18} cm
2D bendover scale	l_{2D}	$0.1\ l_{slab}$
Background magnetic Field	B_0	0.4 nT
Slab/2D ratio	$\delta B_{slab}^2/\delta B_{2D}^2$	0.25
Turbulence strength	$\delta B^2/B_0^2$	0.5

These values should be appropriate for the interstellar medium

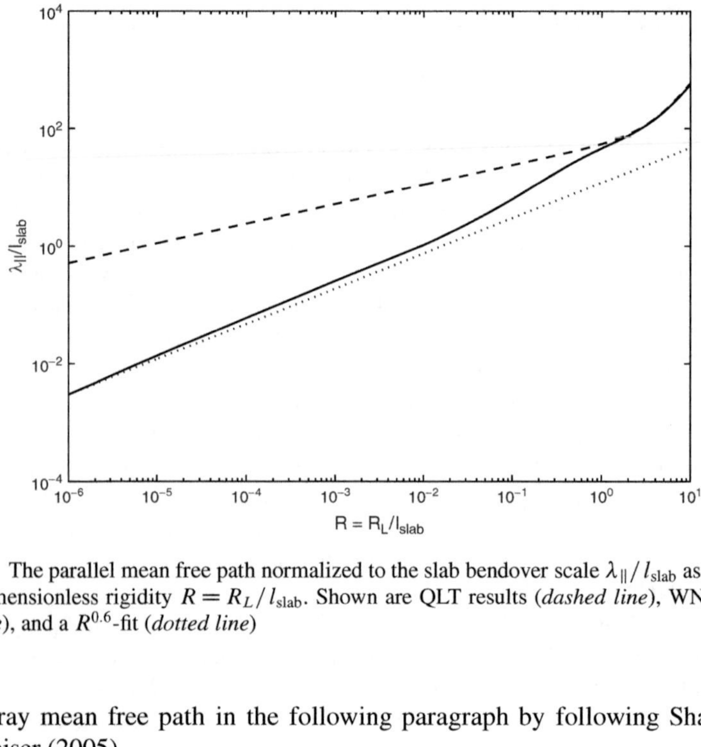

Fig. 8.10 The parallel mean free path normalized to the slab bendover scale $\lambda_{\parallel}/l_{slab}$ as a function of the dimensionless rigidity $R = R_L/l_{slab}$. Shown are QLT results (*dashed line*), WNLT results (*solid line*), and a $R^{0.6}$-fit (*dotted line*)

cosmic ray mean free path in the following paragraph by following Shalchi and Schlickeiser (2005).

8.3.1 *Rigidity Dependence of the Weakly Nonlinear Parallel Mean Free Path*

Figure 8.10 shows the parallel mean free path $\lambda_{\parallel}/l_{slab}$ as a function of the dimensionless rigidity $R = R_L/l_{slab}$, calculated with WNLT in comparison with QLT results. The inverse parallel mean free path $l_{slab}/\lambda_{\parallel}$ is proportional to the primary-to-secondary abundance ratio (see Sect. 1.8) at relativistic rigidities. The dimensionless rigidity R can be related to the kinetic energy E_{kin} by

$$E_{\text{kin}} = \sqrt{E_0^2 + (\alpha R)^2} - E_0 \tag{8.47}$$

with $E_0 = mc^2$ and

$$\alpha = |q| B_0 l_{\text{slab}}. \tag{8.48}$$

Here q is the charge of the particle and B_0 is the background magnetic field (mean field). For such values, the parameter α, defined in (8.48), becomes for electrons and protons $\alpha \approx 2.4 \times 10^6$ GeV. With the definition

$$R_0 := \frac{E_0}{\alpha}, \tag{8.49}$$

we have, for relativistic particles, $R \gg R_0$. For protons, for instance, we find $R_0 \approx 0.4 \times 10^{-6}$. As explained in the next two paragraphs, the results presented here are valid for $R_0 < R < 1$ or in terms of the kinetic energy $E_0 < E_{\text{kin}} < 2.4 \times 10^6$ GeV.

8.3.2 Importance of Nonlinear Effects

The reason for the disagreement between QLT and WNLT is that, for small and medium rigidities, the (nonlinear) 2D pitch-angle Fokker–Planck coefficient controls the parallel mean free path. Within QLT, this coefficient is equal to zero (see Chap. 3). Physically, the steeper rigidity dependence in WNLT arises from the fact that perpendicular diffusion causes resonance broadening. If we take into account this effect, the 2D pitch-angle Fokker–Planck coefficient is no longer equal to zero and, for medium and strong turbulence, the 2D coefficient is even stronger than the slab contribution. Therefore, a coupled theory for parallel and perpendicular diffusion such as WNLT is essential.

In general, the strength of the nonlinear effect is controlled by the ratios $l_{\text{slab}}/l_{2\text{D}}$, $\delta B^2/B_0^2$, and $\delta B_{\text{slab}}^2/\delta B_{2\text{D}}^2$. It is obvious that for slab-like turbulence ($\delta B_{\text{slab}}^2/\delta B_{2\text{D}}^2 \to 0$) or for very weak turbulence ($\delta B^2/B_0^2 \to 0$), quasilinear theory is recovered. More interesting is that we can supress the nonlinear effect also by choosing a small ratio $l_{\text{slab}}/l_{2\text{D}}$ (see Chap. 5). Furthermore, the spectral index 2ν can change the rigidity dependence. For a Kraichnan–Iroshnikov (1965) spectrum, where $2\nu = 3/2$ instead of the Kolmogorov (1941) value of $2\nu = 5/3$, the rigidity dependence of the parallel mean free path would become even steeper than $a = 0.6$.

8.3.3 Validity of the WNLT Results

Because of the simplified turbulence model (dissipationless magnetostatic model) applied here, it is important to discuss the validity of our calculations.Therefore, we split the rigidity range into three parts:

1. The nonrelativistic range $R < R_0$ $(E_{\mathrm{kin}} < E_0)$*:*
Here, dissipation effects, plasma wave effects, and dynamical effects play an essential role. The simple magnetostatic and dissipationless turbulence model considered here is not justified. Therefore, we do not consider nonrelativistic particles in this section.

2. The relativistic range, where $R_0 < R < 1$ $(E_0 < E_{\mathrm{kin}} < 2.4 \times 10^6\,GeV)$*:*
Here, the magnetostatic model should be appropriate. For nonslab models, nonlinear effects cannot be neglected, and, instead of QLT, WNLT must be applied. In this regime, we find a rigidity dependence of $a = 0.6$.

3. The high energy range $R > 1$ $(E_{\mathrm{kin}} > 2.4 \times 10^6\,GeV)$*:*
Here, nonlinear effects should be negligible and quasilinear theory can be applied. In this case, we find sharp gyroresonance with the energy range of the wave spectrum. Because the form of the wave spectrum for such small wavenumbers is not well known, we do not consider such high rigidities. However, we note that we expect a change in the rigidity dependence of the mean free path, which, in principle, could account for the well-established knee in cosmic ray particle spectra.

Because we only consider the relativistic range, where $R < 1$, our results should be valid. The rigidity dependence of $\lambda_{\|}$ must be correct and was confirmed in test-particle simulations (see Chap. 5). Therefore, a value $a = 0.6$ in (8.46) can be explained theoretically by using WNLT calculations. Thus, we conclude that within a nonlinear transport theory, the abundance ratio of primary-to-secondary cosmic ray nuclei can be explained, which thus provides evidence for the nonlinear transport of galactic cosmic rays.

Chapter 9
Summary and Outlook

9.1 Summary

9.1.1 Turbulence and Cosmic Rays

In this book, nonlinear cosmic ray transport theories are discussed. These theories can be used for achieving a more accurate description of plasma–particle interactions in comparison to the standard approach (quasilinear theory). Such interaction can be described by particle diffusion coefficients or mean free paths, and are relevant for describing cosmic ray propagation in different physical systems (e.g., solar system, interstellar space) as well as diffusive shock acceleration at different shock waves (e.g., interplanetary shocks, supernova remnants, active galactic nuclei). To achieve a realistic description of cosmic ray scattering, we have to solve at least three problems:

1. For describing plasma–particle interactions (or wave-particle interaction in the plasma wave picture of turbulence), we have to know the properties of solar wind or interstellar turbulence in the form of the magnetic correlation tensor. Further important properties that have to be known to achieve a realistic description of particle scattering are correlation lengths and field line diffusion coefficients. Such quantities are discussed in Chap. 2.
2. To connect these turbulence properties with particle properties such as diffusion coefficients or mean free paths, we need a formulation such as a first-order perturbation theory (also known as quasilinear theory, see Chap. 3). An alternative possibility for theoretically describing plasma–particle interactions are computer simulations. By comparing quasilinear results with such simulations, it is shown in Chap. 3 that there are at least three problems associated with the application of quasilinear theory, namely the 90°-problem, the problem of perpendicular transport, and the geometry problem. To solve these problems, the quasilinear approximation has to be replaced by a nonlinear theory. Such nonlinear approaches are discussed in Chaps. 4–7. Some examples are the nonlinear guiding center theory (NLGC theory), the weakly nonlinear theory (WNLT), the second-order quasilinear theory (SOQLT), and the extended nonlinear guiding center theory (ENLGC theory).

A. Shalchi, *Nonlinear Cosmic Ray Diffusion Theories*,
Astrophysics and Space Science Library,
DOI: 10.1007/978-3-642-00309-7_9,

3. To obtain a complete description of cosmic ray propagation or acceleration, a formulation is needed for connecting cosmic ray diffusion coefficients with the observable quantities.

As an example, we consider the mechanism of diffusive shock acceleration, which is thought to be responsible for the high cosmic ray energies. If we consider a shock wave that propagates in the direction perpendicular with respect to the mean magnetic field, the acceleration is given by (see, e.g., Webb et al. 1995 or Chap. 8 of this book)

$$\left(\frac{1}{p} \frac{\mathrm{d}p}{\mathrm{d}t} \right)^{-1} = \frac{3 s \kappa_{\perp}}{V_{\mathrm{sh}}^2 (s-1)}. \tag{9.1}$$

On the right-hand side of this equation, we find shock wave properties (s is the shock compression ratio, V_{sh} is the shock velocity) as well as the cosmic ray diffusion coefficient $\kappa_{\perp}$ in the perpendicular direction. By integrating (9.1), one can easily compute the maximum energy a charged particle can get due to the interaction with the shock wave. The maximum energy can be observed. Equation (9.1) provides the connection between this quantity and the diffusion coefficient $\kappa_{\perp}$. To compute the latter parameter, we need a diffusion theory (in the case of perpendicular diffusion, a nonlinear formulation has to be applied) as well as an understanding of the turbulence properties. To provide a realistic description of cosmic ray acceleration at perpendicular propagating interplanetary or interstellar shock waves, we need an improved understanding of turbulence (correlation functions, field line diffusion coefficients) as well as a reliable scattering theory. As demonstrated in this book, the later theory has to be a nonlinear approach.

9.1.2 Specific Conclusions

More specific conclusion that can be drawn from this book are the following:

- In previous linear and nonlinear scattering theories, the assumption that guiding centers follow magnetic field lines has been used. In Chap. 1, we have derived a more general relation ((1.77) and (1.78)) directly from the (exact) Newton–Lorentz equation. This general relation should be the basis for future calculations of the perpendicular mean free path. It will be an interesting matter to explore the parameter regimes for which the guiding center approximation is valid and for which it is not.
- In Chap. 2, an analytical nonlinear theory for field line wandering is presented. Field line random walk is important for understanding cosmic ray scattering in the direction perpendicular to the mean magnetic field. As shown, field line wandering behaves superdiffusively for standard forms of the turbulence wave spectrum. Two conclusions are possible: real field lines in the solar wind and

other systems are not diffusive or the standard wave spectrum has to be replaced by a different form.

- During the past few years, quasilinear theory (QLT) was assumed to be an accurate and reliable tool for describing the interaction between plasma and particle. Therefore, QLT was used to describe parallel and perpendicular spatial diffusion in different physical systems such as the heliosphere or the interstellar medium. It is demonstrated in Chap. 3 that QLT is not always accurate and in some cases invalid. There are at least three problems that occur if QLT is applied:
    - *The 90°-problem:* QLT fails to describe pitch-angle diffusion at 90° ($\mu = 0$). For steep spectra, the 90°-problem causes an infinitely large parallel mean free path $\lambda_{\|}$. For certain improved turbulence models (e.g., plasma wave models, dynamical turbulence model), the singularity can be suppressed. However, QLT results are not reliable due to the failure of the theory itself. By providing a comparison between QLT and simulations, Tautz et al. (2006b) have demonstrated that QLT is not valid also for plasma wave turbulence.
    - *The problem of perpendicular transport:* QLT results are in disagreement with those of the previous test-particle simulations. Within QLT, we find usual (Markovian) diffusion for slab geometry, while we find a strong subdiffusion in test-particle simulations. For nonslab models, such as slab/2D composite model, we find that diffusion is recovered (or nearly recovered). In QLT, however, we find a ballistic motion for all time-scales.
    - *The geometry problem:* For certain turbulence models (e.g., dissipationless slab/2D composite turbulence), 90°-scattering is not important if the parallel mean free path is calculated. Nevertheless, QLT is not very accurate for parallel diffusion in nonslab models as demonstrated by this book. In test-particle simulations, we find a much steeper rigidity dependence than predicted by QLT. Because this problem is related to the turbulence geometry, this third problem of cosmic rays transport is referred to as the *geometry problem*. It should be noted that, for isotropic or anisotropic turbulence, the 90°-problem as well as the geometry problem occur if the parallel mean free path is calculated.

- Some previous theories that have been derived to solve the 90°-problem or the problem of perpendicular transport are discussed in Chap. 4. It is demonstrated that only a few theories can describe pitch-angle diffusion for all values of μ. But these theories are not tractable (e.g., the strong turbulence, weak coupling theory of Goldstein 1976) or they are based on crude models and cannot be derived systematically (e.g., Jones et al. 1973; Völk 1975). These chapters motivate the derivation of a systematic and reliable transport theory to solve the 90°-problem. Also, two theories have been derived to describe perpendicular transport: the Bieber and Matthaeus model and the nonlinear guiding center theory (NLGC theory). Both theories cannot describe subdiffusion in the slab model. Therefore, both theories cannot be applied for slab-like turbulence models. For slab/2D composite geometry, however, NLGC theory agrees very well with test-particle simulations. Therefore, NLGC theory is the basis for developing more advanced transport theories such as the weakly nonlinear theory (WNLT) or the extended nonlinear guiding center theory (ENLGC theory).

- In Chap. 5, the ideas of the NLGC theory are combined with QLT. The resulting theory, the weakly nonlinear theory (WNLT), can solve the geometry problem as well as the problem of perpendicular transport. Because of the analytical character of this improved approach, the reason for the geometry problem can be understood: perpendicular diffusion causes resonance broadening, which causes a large 2D pitch-angle Fokker–Planck coefficient. For small and medium rigidities, $D^{2D}_{\mu\mu}$ dominates the slab contribution. Obviously, parallel diffusion (or pitch-angle diffusion) and perpendicular diffusion are coupled, and a linear description is not appropriate. From QLT formulas for the different Fokker–Planck coefficients, we can deduce the corresponding WNLT results by the formal substitution of resonance functions:

$$R_n^{\mathrm{QLT}}(\mathbf{k}) \to R_n^{\mathrm{WNLT}}(\mathbf{k}) = \frac{D_\perp k_\perp^2 + \omega}{\left(D_\perp k_\perp^2 + \omega\right)^2 + \left(k_\| v_\| + n\Omega\right)^2}, \tag{9.2}$$

with

$$\omega = \begin{cases} \frac{2D_{\mu\mu}}{1-\mu^2} & \text{for perpendicular diffusion} \\ 0 & \text{for pitch-angle diffusion.} \end{cases} \tag{9.3}$$

- Although WNLT can successfully solve the geometry problem and the problem of perpendicular transport, the weakly nonlinear approach cannot solve the 90°-problem. In Chap. 6, a systematic derivation of a second-order Fokker–Planck coefficient is presented. By comparing second-order results with previous simulations, we find that second-order quasilinear theory (SOQLT) is a reliable tool to describe pitch-angle diffusion close to 90°. Within SOQLT, we find the following form of the resonance function

$$R_n^{\mathrm{SOQLT}}(k_\|) = \int_0^\infty \mathrm{d}t\, \cos\left[\left(k_\| v\mu + n\Omega\right)t\right] \mathrm{e}^{-v^2 k_\|^2 \int_0^t \mathrm{d}\tau \left[\tau^2 D_{\mu\mu}^{(1)}(\tau)\right]}, \tag{9.4}$$

where $D_{\mu\mu}^{(1)}$ is the time-dependent quasilinear pitch-angle Fokker–Planck coefficient.

- WNLT as well as standard NLGC theory agree with test-particle simulations for composite geometry. But, as soon as the two-component model is replaced by the slab model, both theories fail: they cannot describe the subdiffusive behavior that is seen in the simulations. In Chap. 7, an improved (or extended) transport theory is developed that uses less assumptions then the original NGLC theory (or WNLT). By applying an improved treatment of the slab contribution, we can describe subdiffusion for slab geometry as well as the recovery of diffusion for nonslab models. We call this approach the extended nonlinear guiding center (ENLGC) theory. The same improvement or modification can also be applied onto WNLT. Within the ENLGC theory, we find for slab geometry (subdiffusion):

$$\lambda_{\perp} = 6C(\nu)\frac{\delta B_{\mathrm{slab}}^2}{B_0^2}\sqrt{\frac{\pi\lambda_{\parallel} l_{\mathrm{slab}}}{3\tau}} \sim t^{-1/2}, \tag{9.5}$$

and for slab/2D composite geometry (recovery of diffusion)

$$\lambda_{\perp} = \begin{cases} \frac{1}{2}\frac{\delta B_{\mathrm{2D}}^2}{B_0^2}\lambda_{\parallel} & \text{if} \quad \lambda_{\parallel}\lambda_{\perp} \ll 3l_{\mathrm{2D}}^2 \\ \left(\sqrt{3}\pi C(\nu) l_{\mathrm{2D}}\frac{\delta B_{\mathrm{2D}}^2}{B_0^2}\right)^{2/3}\lambda_{\parallel}^{1/3} & \text{if} \quad \lambda_{\parallel}\lambda_{\perp} \gg 3l_{\mathrm{2D}}^2. \end{cases} \tag{9.6}$$

This result is valid for the used wave spectrum. For other spectra, one expects a different form of the perpendicular mean free path.

- In Sect. 8.1, it is demonstrated that QLT for parallel transport and (standard) NLGC theory for perpendicular diffusion can describe measured mean free paths in the solar wind if the magnetostatic turbulence model is replaced by an improved dynamical turbulence model, namely the nonlinear anisotropic dynamical turbulence model (NADT model). These new results show indications that QLT could be recovered for parallel diffusion if there are strong dynamical effects. It should be the subject of future work to explore in detail the possible recovery of QLT for certain turbulence models. In Sect. 8.2, the ENLGC-theory is applied for computing the acceleration timescale of charged particles at highly perpendicular interplanetary shocks. As shown, the nonlinear result provides a much shorter acceleration timescale as the result obtained by using the Bohm limit for the perpendicular diffusion coefficient. This implies rapid acceleration at perpendicular shocks and gives, therefore, rise to the assumption that nonlinear effects in cosmic ray diffusion theory are also important for particle acceleration at shock waves. In Sect. 8.3, it is shown that WNLT can explain the observed abundance ratio of primary-to-secondary cosmic rays. The WNLT result, $\lambda_{\parallel} \sim R^{0.6}$, is in accordance with observations, which thus provides evidence for the nonlinear transport of galactic cosmic rays.

9.2 Outlook

This work, which is mainly of theoretical nature, describes how the different problems of cosmic ray transport theory can be solved. Although the theories described herein provide useful tools to calculate parallel and perpendicular diffusion coefficients, this book can also be seen as a motivation for several future projects.

9.2.1 Future Test-particle Simulations

Because of the new results obtained in this work, the following new research related to test-particle simulations should be performed:

- An improved wave spectrum in the dissipation range should be applied to investigate low energy cosmic rays numerically
- Electric fields should be included to explore the importance of these fields onto spatial diffusion. Furthermore, a numerical investigation of cosmic ray momentum diffusion would be possible.
- Dynamical turbulence effects should be included. Only then the question whether we really can reproduce heliospheric observations by using dynamical turbulence models such as the NADT model can be answered.
- The coupled system of turbulence and charged particles should be simulated.

9.2.2 Future Theoretical Work

This book motivates also some theoretical projects. Some examples are as follows:

- All theories proposed here to solve the problem of perpendicular transport are based on the assumption that guiding centers follow magnetic field lines ($\tilde{v}_x = v_z \delta B_x / B_0$). In Chap. 1, however, a more generalized equation for the guiding center velocity has been derived. Although the application of this improved relation is more complicated, a calculation of linear and nonlinear diffusion coefficients should be possible. It should be the subject of future work to combine the ENLGC theory and the WNLT with the improved guiding center equation and to answer the question under what conditions the previous results can be recovered.
- NLGC theory as well as WNLT are based on several models such as the exponential velocity correlation function. ENLGC theory does not use some of these models, but the improvement of the theory can only be applied for slab geometry. How can a more systematic and reliable theory be formulated for nonslab turbulence?
- Several theories are discussed to solve the different problems. SOQLT, for instance, can solve the 90°-problem, whereas WNLT or ENLGC theory can describe perpendicular diffusion and can solve the geometry problem. A unified transport theory that can solve all problems related to cosmic ray transport theory has to be developed.
- Most calculations presented here have been performed for magnetostatic turbulence. Although not much is known from measurements about the one-point-two-time correlation function, more realistic models are available (e.g., the NADT model). A combination of such improved turbulence models with nonlinear transport theories has to be achieved in the near future. Then, also the question can be answered whether QLT is recovered for parallel scattering in dynamical turbulence as discussed in Chap. 8.

- The SOQLT, which was derived to solve the 90°-problem, provides a reasonable agreement with the test-particle simulations (see Chap. 6) for slab geometry. However, 90°-scattering is also essential for isotropic and anisotropic turbulence (see Tautz et al. 2006a, b). Therefore, SOQLT has to be reformulated for isotropic turbulence, and a comparison between these new results and previous test-particle simulations should be provided.
- All previous transport theories assumed axisymmetric turbulence. How do diffusion coefficients in a nonsymmetric systems look like? The answer for this question is important and interesting because real physical systems are not symmetric.
- Three examples for the application of nonlinear transport theories are discussed in Chap. 8. It is demonstrated that a nonlinear treatment of particle orbits is essential to understand the measured perpendicular mean free paths in the heliosphere. Also shown is the influence of nonlinear effects on particle acceleration at perpendicular interplanetary shock waves. Furthermore, we need a nonlinear formulation of parallel transport to explain the observed abundance ratio of primary-to-secondary cosmic rays. The existing theories such as WNLT, SOQLT, and ENLGC theory have to be applied on other physical situations to explore the importance of nonlinear effects. Further examples are diffusive shock acceleration at supernova shock waves or active galactic nuclei.

9.2.3 Future Observational Work

In this book, different theories are discussed. Most of them are compared with simulations and observations. A main result of this book is that several parameters are key input parameters in the calculation of cosmic ray diffusion coefficients. In previous articles, these parameters have been assumed to be unimportant. Therefore, this work also motivates future observations:

- By applying WNLT and QLT and by employing the NADT model, it is demonstrated in this book that the ratio l_{2D}/l_{slab} has a strong influence on dynamical and nonlinear effects. For instance, QLT can be recovered for larger values of l_{2D}/l_{slab}. There are indications from laboratory experiments such as Robinson and Rusbridge (1971) that the parallel correlation scale may be much larger (about ten times) than the perpendicular scale in an evolving magnetofluid plasma. In astrophysical systems, this ratio is unclear, but the assumption $l_{2D}/l_{slab} = 0.1$ was used in previous papers (e.g., Matthaeus et al. 2003) and in some calculations presented in this book. Furthermore, it is demonstrated that if $l_{2D}/l_{slab} = 0.1$ is true for solar wind parameters, nonlinear effects are essential for parallel diffusion. Therefore, an accurate measurement of this ratio should be the subject of future work.
- At least in Chap. 8, but also in previous articles (e.g., Bieber et al. 1994), it has been demonstrated that dynamical turbulence effects are important if parallel and perpendicular mean free paths are calculated. The time-dependence must

be included into transport theory calculations in the form of dynamical correlation function. A measurement of this function in the form of one-point-two-time correlation functions $\langle \delta B_i(0,t) \delta B_j(0,0) \rangle$ should be the subject of future work.

- By applying WNLT, we deduce a steep rigidity dependence of the parallel mean free path. At least for medium rigidities ($0.01 \leq R \leq 1$), WNLT can be confirmed by using a test-particle code. Furthermore, the magnetostatic model should be appropriate for such particles because the restriction $v \gg v_A$ is fulfilled, and dynamical effects as well as electric field effects can, therefore, be neglected. Thus, the magnetostatic WNLT calculations presented in this work are reliable. Therefore, one would expect that the rigidity dependence is indeed steeper ($\lambda_\parallel \sim R^{0.6}$) even for solar wind parameters. It should be the subject of future work to find observational indications for the steeper rigidity dependence in the interplanetary system.

References

Abramowitz, M., Stegun, I.A.: Handbook of Mathematical Functions. Dover, New York (1974)

Axford, W.I., Leer, E., Skadron, G.: The acceleration of cosmic ray by shock waves. Proceedings of the International Conference on Cosmic Rays 15th, Akademiia na Naukite, B'lgarska, **11**, 132 (1977)

Balescu, R., Wang, H.-D., Misguich, J.H.: Langevin equation versus kinetic equation: subdiffusive behavior of charged particles in a stochastic magnetic field. Phys. Plasmas **1**, 3826 (1994)

Batchelor, G.K.: The Theory of Homogeneous Turbulence. Cambridge University Press, Cambridge (1953)

Bavassano, B.: Observations of Alfvénic turbulence evolution in the 3D heliosphere. AIP Conference Proceedings **679**, 377 (2003)

Bavassano Cattaneo, M.B., Formisano, V.: Particle acceleration at comets. Phys. Scr. T **T50**, 98 (1994)

Beck, R., Berkhuijsen, E.M., Uyaniker, B.: Studies of turbulent interstellar magnetic fields using radio emission. In: M. Ostrowski, R. Schlickeiser (eds.), Plasma Turbulence and Energetic Particles in Astrophysics, Krakow (1999)

Bell, A.R.: The acceleration of cosmic rays in shock fronts I. Mon. Not. R. Astron. Soc. **182**, 147 (1978a)

Bell, A.R.: The acceleration of cosmic rays in shock fronts II. Mon. Not. R. Astron. Soc. **182**, 443 (1978b)

Ben-Israel, I., Piran, T., Eviatar, A., Weinstock, J.: A statistical theory of electromagnetic waves in turbulent plasmas. Astrophys. Space Sci. **38**, 125 (1975)

Berezhko, E.G., Völk, H.J.: Spectrum of cosmic rays produced in supernova remnants. Astrophys. J. **661**, L175 (2007)

Berezhko, E.G., Ksenofontov, L.T., Völk, H.J.: Emission of SN 1006 produced by accelerated cosmic rays. Astron. Astrophys. **395**, 943 (2002)

Bieber, J.W., Matthaeus, W.H.: Cosmic Ray Pitch Angle Scattering in Dynamical Magnetic Turbulence. Proceedings of the 22nd International Cosmic Ray Conference. 11–23 August, 1991. Dublin, Ireland **3**, 248 (1991)

Bieber, J.W., Matthaeus, W.H.: Particle transport from a turbulence perspective. AIP Conference Proceedings **264**, 86 (1992)

Bieber, J.W., Matthaeus, W.H.: Perpendicular diffusion and drift at intermediate cosmic-ray energies. Astrophys. J. **485**, 655 (1997)

Bieber, J.W., Earl, J.A., Green, G., Kunow, H., Mueller-Mellin, R., Wibberenz, G.: Interplanetary pitch angle scattering and coronal transport of solar energetic particles – new information from HELIOS. J. Geophys. Res. **85**, 2313 (1980)

Bieber, J.W, Smith, C.W., Matthaeus, W.H.: Cosmic-ray pitch-angle scattering in isotropic turbulence. Astrophys. J. **334**, 470 (1988)

Bieber, J.W., Matthaeus, W.H., Smith, C.W., Wanner, W., Kallenrode, M.-B., Wibberenz, G.: Proton and electron mean free paths: the Palmer consensus revisited. Astrophys. J. **420**, 294 (1994)

Bieber, J.W., Wanner, W., Matthaeus, W.H.: Dominant two-dimensional solar wind turbulence with implications for cosmic ray transport. J. Geophys. Res. **101**, 2511 (1996)

Binns, W.R., Fickle, R.K., Waddington, C.J., Garrard, T.L., Stone, E.C., Israel, M.H., Klarmann, J.: Cosmic-ray abundances of elements with atomic number 26 less than or equal to 40 measured on HEAO 3. Astrophys. J. **247**, L115 (1981)

Blandford, R.D., Ostriker, J.P.: Particle acceleration by astrophysical shocks. Astrophys. J. **221**, 269 (1978)

Burger, R.A., Hattingh, M.: Toward a realistic diffusion tensor for galactic cosmic rays. Astrophys. J. **505**, 244 (1998)

Burger, R.A., Potgieter, M.S., Heber, B.: Rigidity dependence of cosmic ray proton latitudinal gradients measured by the Ulysses spacecraft: implications for the diffusion tensor. J. Geophys. Res. **105**, 27447 (2000)

Candia, J., Roulet, E.: Diffusion and drift of cosmic rays in highly turbulent magnetic fields. J. Cosmol. Astropart. Phys. **10**, 007 (2004)

Chenette, D.L.: The propagation of Jovian electrons to earth. J. Geophys. Res. **85**, 2243 (1980)

Chenette, D.L., Conlon, T.F., Pyle, K.R., Simpson, J.A.: Observations of Jovian electrons at 1 AU throughout the 13 month Jovian synodic year. Astrophys. J. **215**, L95 (1977)

Corrsin, S.: Progress report on some turbulent diffusion research. In: F. Frenkiel, P. Sheppard (eds.), Atmospheric Diffusion and Air Pollution, Advanced in Geophysics, Vol. 6, Academic, New York (1959)

Debrunner, H., Lockwood, J.A.: The spatial anisotropy, rigidity spectrum, and propagation characteristics of the relativistic solar particles during the event on May 7, 1978. J. Geophys. Res. **85**, 6853 (1980)

Denskat, K.U., Neubauer, F.M.: Observations of hydrodynamik turbulence in the solar wind. In: M. Neugebauer (ed.),Solar Wind Five, Proceedings of a conference held in Woodstock, Vermont, 1–5 November, 1982, vol. 2280 of NASA Conference Publication, pp. 81–91, NASA, Washington, USA (1982)

Desai, M.I., Mason, G.M., Dwyer, J.R., Mazur, J.E., Gold, R.E., Krimigis, S.M., Smith, C.W., Skoug, R.M.: Evidence for a suprathermal seed population of heavy ions accelerated by interplanetary shocks near 1 AU. Astrophys. J. **588**, 1149 (2003)

Dröge, W.: Particle scattering by magnetic fields. Space Sci. Rev. **93**, 121 (2000)

Dupree, T.H.: A perturbation theory of strong plasma turbulence. Phys. Fluids **9**, 1773 (1966)

Dupree, T.H.: Nonlinear theory of drift-wave turbulence and enhanced diffusion. Phys. Fluids **10**, 1049 (1967)

Dwyer, J.R., Mason, G.M., Mazur, J.E., Jokipii, J.R., von Rosenvinge, T.T., Lepping, R.P.: Perpendicular transport of low-energy corotating interaction region-associated nuclei. Astrophys. J. **490**, L115 (1997)

Earl, J.A.: Diffusion of charged particles in a random magnetic field. Astrophys. J. **180**, 227 (1973)

Earl, J.A.: The diffusive idealization of charged-particle transport in random magnetic fields. Astrophys. J. **193**, 231 (1974)

Ferrando, P.: MeV to GeV electron propagation and modulation: results of the KET-telescope onboard ULYSSES. Adv. Space Res. **19**, 905 (1997)

Ferreira, S.E.S., Potgieter, M.S., Burger, R.A., Heber, B., Fichtner, H., Lopate, C.: Modulation of Jovian and galactic electrons in the heliosphere: Part 2. Radial transport of a few MeV electrons. J. Geophys. Res. **106**, 29313 (2001)

Fichtner, H.: Anomalous Cosmic Rays: messengers from the Outer Heliosphere. *Habilitationsschrift*, Rheinische Friedrich-Wilhelms-Universität zu Bonn, Germany (2000)

Fichtner, H.: Anomalous cosmic rays: messengers from the outer heliosphere. Space Sci. Rev. **95**, 639 (2001)

Fisk, L.A., Goldstein, M.L., Klimas, A.J., Sandri, G.: The Fokker–Planck coefficient for pitch-angle scattering of cosmic rays. Astrophys. J. **190**, 417 (1974)

Fletcher, L.: Numerical simulations of coronal particle trapping. Astron. Astrophys. **326**, 1259 (1997)

Ford, T., Palmer, I.D., Sanders, R.: Separation of solar and interplanetary diffusion in solar cosmic ray events. J. Geophys. Res. **82**, 4704 (1977)

Giacalone, J., Jokipii, J.R.: The transport of cosmic rays across a turbulent magnetic field. Astrophys. J. **520**, 204 (1999)

Gleeson, L.: The equations describing the cosmic-ray gas in the interplanetary region. Planet. Space Sci. **17**, 31 (1969)

Gloeckler, G., Schwadron, N.A., Fisk, L.A., Geiss, J.: Weak pitch angle scattering of few MV rigidity ions from measurments of anisotropies in the distribution function of interstellar pickup H+. Geophys. Res. Lett. **22**, 2665 (1995)

Goldreich, P., Sridhar, S.: Toward a theory of interstellar turbulence. Part 2: Strong alfvénic turbulence. Astrophys. J. **438**, 763 (1995)

Goldstein, M.L.: A nonlinear theory of cosmic-ray pitch-angle diffusion in homogeneous magnetostatic turbulence. Astrophys. J. **204**, 900 (1976)

Gordon, B.E., Lee, A.M., Möbius, E., Trattner, K.J.: Coupled hydromagnetic wave excitation and ion acceleration at interplanetary traveling shocks and Earth's bow shock revisited. J. Geophys. Res. **104**, 28263 (1999)

Gradshteyn, I.S., Ryzhik, I.M.: Table of Integrals, Series, and Products. Academic Press, New York (2000)

Green, M.S.: Brownian motion in a gas of noninteracting molecules. J. Chem. Phys. **19**, 1036 (1951)

Hamilton, D.C.: The radial transport of energetic solar flare particles from 1 to 6 AU. J. Geophys. Res. **82**, 2157 (1977)

Hasselmann, K., Wibberenz, G.: Scattering of charged particles by random electromagnetic fields. Z. Geophys. **34**, 353 (1968)

Honda, M., Honda, Y.S.: Filamentary jets as a cosmic-ray "Zevatron." Astrophys. J. **617**, L37 (2004)

Jaekel, U., Schlickeiser, R.: The Fokker–Planck coefficients of cosmic ray transport in random electromagnetic fields. J. Phys. G **18**, 1089 (1992)

Jokipii, J.R.: Cosmic-ray propagation. I. Charged particles in a random magnetic field. Astrophys. J. **146**, 480 (1966)

Jokipii, J.R.: Addendum and erratum to cosmic-ray propagation. I. Astrophys. J. **152**, 671 (1968)

Jokipii, J.R.: Rate of energy gain and maximum energy in diffusive shock acceleration. Astrophys. J. **313**, 842 (1987)

Jokipii, J.R., Parker, E.N.: Cosmic-ray life and the stochastic nature of the galactic magnetic field. Astrophys. J. **155**, 799 (1969)

Jokipii, J.R., Levy, E.H., Hubbard, W.B.: Effects of particle drift on cosmic-ray transport. I. General properties, application to solar modulation. Astrophys. J. **213**, 861 (1977)

Jokipii, J.R., Kóta, J., Giacalone, J.: Perpendicular transport in one- and two-dimensional shock simulations. Geophys. Res. Lett. **20**, 1759 (1993)

Jokipii, J.R., Kóta, J., Giacalone, J., Horbury, T.S., Smith, E.J.: Interpretation and consequences of large-scale magnetic variances observed at high heliographic latitude. Geophys. Res. Lett. **22**, 3385 (1995)

Jones, F.C., Kaiser, T.B., Birmingham, T.J.: New approach to cosmic-ray diffusion theory. Phys. Rev. Lett. **31**, 485 (1973)

Jones, F.C., Birmingham, T.J., Kaiser, T.B.: Partially averaged field approach to cosmic ray diffusion. Phys. Fluids **21**, 347 (1978)

Jones, F.C., Lukasiak, A., Ptuskin, V., Webber, W.: The modified weighted slab technique: models and results. Astrophys. J. **547**, 264 (2001)

Kaiser, T.B.: Simulation of the velocity diffusion of charged particles in turbulent magnetic fields. In: International Cosmic Ray Conference, 14th, Munich, West Germany, August 15–29, Conference Papers. **3**, 86 (1975)

Kaiser, T.B., Jones, F.C., Birmingham, T.J.: Cosmic rays in a random magnetic field: breakdown of the quasilinear derivation of the kinetic equation. Astrophys. J. **180**, 239 (1973)

Kaiser, T.B., Birmingham, T.J., Jones, F.C.: Computer simulation of the velocity diffusion of cosmic rays. Phys. Fluids **21**, 361 (1978)

Klimas, A.J., Sandri, G.: Foundation of the theory of cosmic-ray transport in random magnetic fields. Astrophys. J. **169**, 41 (1971)

Klimas, A.J., Sandri, G.: A rigorous cosmic-ray transport equation with no restrictions on particle energy. Astrophys. J. **180**, 937 (1973)

Kolmogorov, A.N.: The local structure of turbulence in incompressible viscous fluid for very large Reynolds number. Dokl. Akad. Nauk SSSR **30**, 301 (1941)

Kota, J., Jokipii, J.R., Kopriva, D.A., Gombosi, T.I., Owens, A.J., Merenyi, E.: A numerical study of the pitch-angle scattering of cosmic rays. Astrophys. J. **254**, 398 (1982)

Kóta, J., Jokipii, J.R.: Velocity correlation and the spatial diffusion coefficients of cosmic rays: compound diffusion. Astrophys. J. **531**, 1067 (2000)

Kraichnan, R.H.: Inertial-range spectrum of hydromagnetic turbulence. Phys. Fluids **8**, 1385 (1965)

Krymsky, G.F.: A regular mechanism for the acceleration of charged particles on the front of a shock wave. Dokl. Akad. Nauk. SSSR **234**, 1306 (1977)

Kubo, R.: Statistical-mechanical theory of irreversible processes. I. General theory and simple applications to magnetic and conduction problems. J. Phys. Soc. Jpn. **12**, 570 (1957)

Kubo, R.: Generalized cumulant expansion method. J. Phys. Soc. Jpn. **17**, 1100 (1962)

Lazar, M., Spanier, F., Schlickeiser, R.: Linear damping and energy dissipation of shear Alfvén waves in the interstellar medium. Astron. Astrophys. **410**, 415 (2003)

le Roux, J.A., Webb, G.M.: Nonlinear cosmic-ray diffusive transport in combined two-dimensional and slab magnetohydrodynamic turbulence: a BGK–Boltzmann approach. Astrophys. J. **667**, 930 (2007)

Lee, M.A.: Coupled hydromagnetic wave excitation and ion acceleration at interplanetary traveling shocks. J. Geophys. Res. **88**, 6109 (1983)

Lerche, I., Schlickeiser, R.: Cosmic ray transport in anisotropic magnetohydrodynamic turbulence. I. Fast magnetosonic waves. Astron. Astrophys. **378**, 279 (2001)

Li, G., Zank, G.P.: Mixed particle acceleration at CME-driven shocks and flares. Geophys. Res. Lett. **32**, L02101 (2005)

Li, G., Zank, G.P., Rice, W.K.M.: Energetic particle acceleration and transport at coronal mass ejection drive shocks. J. Geophys. Res. **108**, 1082 (2003)

Li, G., Zank, G.P., Rice, W.K.M.: Acceleration and transport of heavy ions at coronal mass ejection-driven shocks. J. Geophys. Res. **110**, A06104 (2005)

Lin, R.P.: Observations of scatter-free propagation of approximately 40-keV solar electrons in the interplanetary medium. J. Geophys. Res. **75**, 2583 (1970)

Lin, R.P.: Non-relativistic solar electrons. Space Sci. Rev. **16**, 189 (1974)

Lockwood, J.A., Debrunner, H., Flueckiger, E., Neuenschwander, H., Schubnell, M.: A model of propagation of solar flare particles and its application to cosmic ray ground level events. J. Geophys. Res. **87**, 4338 (1982)

Ma Sung, L.S., Earl, J.A.: Interplanetary propagation of flare-associated energetic particles. Astrophys. J. **222**, 1080 (1978)

Mace, J.W., Matthaeus, W.H., Bieber, J.W.: Numerical investigation of perpendicular diffusion of charged test particles in weak magnetostatic slab turbulence. Astrophys. J. **538**, 192 (2000)

Matthaeus, W.H., Goldstein, M.L.: Stationarity of magnetohydrodynamic fluctuations in the solar wind. J. Geophys. Res. **87**, 10347 (1982)

Matthaeus, W.H., Smith, C.: Structure of correlation tensors in homogeneous anisotropic turbulence. Phys. Rev. A **24**, 2135 (1981)

Matthaeus, W.H., Goldstein, M., Roberts, D.A.: Evidence for the presence of quasi-two-dimensional nearly incompressible fluctuations in the solar wind. J. Geophys. Res. **95**, 20673 (1990)

Matthaeus, W.H., Gray, P.C., Pontius, D.H., Jr., Bieber, J.W.: Spatial structure and field-line diffusion in transverse magnetic turbulence. Phys. Rev. Lett. **75**, 2136 (1995)

Matthaeus, W.H., Qin, G., Bieber, J.W., Zank, G.P.: Nonlinear collisionless perpendicular diffusion of charged particles. Astrophys. J. **590**, L53 (2003)

Matthaeus, W.H., Bieber, J.W., Ruffolo, D., Chuychai, P., Minnie, J.: Spectral properties and length scales of two-dimensional magnetic field models. Astrophys. J. **667**, 956 (2007)

McCarthy, J., O'Gallagher, J.J.: The radial variation of solar flare proton anisotropies observed in deep space on Pioneers 10 and 11. Geophys. Res. Lett. **3**, 53 (1976)

McComb, W.D.: The Physics of Fluid Turbulence. Oxford University Press, New York, p 594 (1990)

Michalek, G.: On the cosmic ray cross-field diffusion for solar wind turbulence. Astron. Astrophys. **376**, 667 (2001)

Michalek, G., Ostrowski, M.: Cosmic ray momentum diffusion in the presence of nonlinear Alfvén waves. Nonlin. Processes Geophys. **3**, 66 (1996)

Minnie, J.: Observational Constraints on the Heliospheric Diffusion Tensor for Galactic Cosmic Rays. Master Thesis, University of Potchefstroom, South Africa (2002)

Minter, A.H., Spangler, S.R.: Observation of turbulent fluctuations in the interstellar plasma density and magnetic field on spatial scales of 0.01–100 Parsecs. Astrophys. J. **458**, 194 (1996)

Möbius, E., Rucinski, D., Lee, M.A., Isenberg, P.A.: Decreases in the antisunward flux of interstellar pickup He+ associated with radial interplanetary magnetic field. J. Geophys. Res. **10**, 257 (1998)

Oughton, S., Priest, E., Matthaeus W.H.: The influence of a mean magnetic field on three-dimensional magnetohydrodynamic turbulence. J. Fluid Mech. **280**, 95 (1994)

Oughton, S., Dmitruk, P., Matthaeus, W.H.: A two-component phenomenology for homogeneous magnetohydrodynamic turbulence. Phys. Plasmas **13**, 042306 (2006)

Owens, A.J.: Cosmic-Ray Scintillations. Dissertation Abstracts International, 34–07, B, 3423 (1973)

Owens, A.J.: The effects of nonlinear terms in cosmic-ray diffusion theory. Astrophys. J. **191**, 235 (1974)

Owens, A.J., Gombosi, T.: Cosmic-ray scattering in simulated interplanetary magnetic field fluctuations. Astrophys. J. **235**, 1071 (1980)

Palmer, I.D.: Transport coefficients of low-energy cosmic rays in interplanetary space. Rev. Geophys. Space Phys. **20**, 335 (1982)

Palmer, I.D., Palmeira, R.A.R., Allum, F.R.: Monte Carlo model of the highly anisotropic solar proton event of 20 April 1971. Solar Phys. **40**, 449 (1975)

Palmer, I.D., Zwickl, R.D., Webber, W.R., McDonald, F.B.: Spectrum of mean free paths near earth in the solar cosmic ray event of April 29, 1973. J. Geophys. Res. **83**, 2461 (1978)

Potter, D.W., Lin, R.P., Anderson, K.A.: Impulsive 2–10 keV solar electron events not associated with flares. Astrophys. J. **236**, L97 (1980)

Press, W.H., Flannery, B.P., Teukolsky, S.A., Vetterling, W.T.: Numerical Recipes. Cambridge University Press, Cambridge (1986)

Ptuskin, V.S.: Propagation, confinement models, and large-scale dynamical effects of galactic cosmic rays. Space Sci. Rev. **99**, 281 (2001)

Qin, G.: Charged Particle Transport in Magnetic Fields Turbulence and Study of Trim Simulation and SSX Experiment. Ph.D. Thesis, University of Delaware, USA (2002)

Qin, G., Matthaeus, W.H., Bieber, J.W.: Subdiffusive transport of charged particles perpendicular to the large scale magnetic field. Geophys. Res. Lett. **29**, 1048 (2002a)

Qin, G., Matthaeus, W.H., Bieber, J.W.: Perpendicular transport of charged particles in composite model turbulence: recovery of diffusion. Astrophys. J. **578**, L117 (2002b)

Qin, G., Matthaeus, W.H., Bieber, J.W.: Parallel diffusion of charged particles in strong two-dimensional turbulence. Astrophys. J. **640**, L103 (2006)

Reames, D.V.: Particle acceleration at the Sun and in the heliosphere. Space Sci. Rev. **90**, 413 (1999)

Rice, W.K.M., Zank, G.P., Li, G.: Particle acceleration and coronal mass ejection driven shocks: shocks of arbitrary strength. J. Geophys. Res. **108**, 1369 (2003)

Robinson, D.C., Rusbridge, M.G.: Structure of turbulence in the Zeta plasma. Phys. Fluids **14**, 2499 (1971)
Ruffolo, D., Matthaeus, W.H., Chuychai, P.: Separation of magnetic field lines in two-component turbulence. Astrophys. J. **614**, 420 (2004)
Ruffolo, D., Chuychai, P., Matthaeus, W.H.: Random walk of magnetic field lines in nonaxisymmetric turbulence. Astrophys. J. **644**, 971 (2006)
Salu, Y., Montgomery, D.C.: Turbulent diffusion from a quasi-kinematical point of view. Phys. Fluids **20**, 1 (1977)
Schlickeiser, R.: Cosmic Ray Astrophysics. Springer, Berlin (2002)
Schlickeiser, R., Achatz, U.: Cosmic-ray particle transport in weakly turbulent plasmas. I: Theory. J. Plasma Phys. **49**, 63 (1993)
Schulze, B.M., Richter, A.K., Wibberenz, G.: Influence of finite injections and of interplanetary propagation on time-intensity and time-anisotropy profiles of solar cosmic rays. Solar Phys. **54**, 207 (1977)
Shalchi, A.: Second-order quasilinear theory of cosmic ray transport. Phys. Plasmas **12**, 052905 (2005a)
Shalchi, A.: Time-dependent transport and subdiffusion of cosmic rays. J. Geophys. Res. **110**, A09103 (2005b)
Shalchi, A.: Analytical investigation of the two-dimensional cosmic ray Fokker–Planck equation. Astron. Astrophys. **448**, 809 (2006a)
Shalchi, A.: Extended nonlinear guiding center theory of perpendicular transport. Astron. Astrophys. **453**, L43 (2006b)
Shalchi, A., Döring, H.: Velocity correlation functions of charged test particles. J. Phys. G **34**, 859 (2007)
Shalchi, A., Kourakis, I.: Analytical description of stochastic field-line wandering in magnetic turbulence. Phys. Plasmas **14**, 092903 (2007)
Shalchi, A., Schlickeiser, R.: The parallel mean free path of heliospheric cosmic rays in composite slab/two-dimensional geometry. I. The damping model of dynamical turbulence. Astrophys. J. **604**, 861 (2004a)
Shalchi, A., Schlickeiser, R.: Quasilinear perpendicular diffusion of cosmic rays in weak dynamical turbulence. Astron. Astrophys. **420**, 821 (2004b)
Shalchi, A., Schlickeiser, R.: Evidence for the nonlinear transport of galactic cosmic rays. Astrophys. J. **626**, L97 (2005)
Shalchi, A., Bieber, J.W., Matthaeus, W.H.: Analytic forms of the perpendicular diffusion coefficient in magnetostatic turbulence. Astrophys. J. **604**, 675 (2004a)
Shalchi, A., Bieber, J.W., Matthaeus, W.H., Qin, G.: Nonlinear parallel and perpendicular diffusion of charged cosmic rays in weak turbulence. Astrophys. J. **616**, 617 (2004b)
Shalchi, A., Bieber, J.W., Matthaeus, W.H.: Nonlinear guiding center theory of perpendicular diffusion in dynamical turbulence. Astrophys. J. **615**, 805 (2004c)
Shalchi, A., Bieber, J.W., Matthaeus, W.H., Schlickeiser, R.: Parallel and perpendicular transport of heliospheric cosmic rays in an improved dynamical turbulence model. Astrophys. J. **642**, 230 (2006)
Shalchi, A., Kourakis, I., Dosch, A.: Generalized compound transport of charged particles in turbulent magnetized plasmas. J. Phys. A **40**, 11191 (2007)
Shebalin, J.V., Matthaeus, W.H., Montgomery, D.: Anisotropy in MHD turbulence due to a mean magnetic field. J. Plasma Phys. **29**, 525 (1983)
Spangler, S.R.: The dissipation of magnetohydrodynamic turbulence responsible for interstellar scintillation and the heating of the interstellar medium. Astrophys. J. **376**, 540 (1991)
Spanier, F., Schlickeiser, R.: Damping and wave energy dissipation in the interstellar medium. II. Fast magnetosonic waves. Astron. Astrophys. **436**, 9 (2005)
Swordy, S.P., Müller, D., Meyer, P., L'Heureux, J., Grunsfeld, J.M.: Relative abundances of secondary and primary cosmic rays at high energies. Astrophys. J. **349**, 625 (1990)

Tautz, R.C., Shalchi, A., Schlickeiser, R.: Comparison between test-particle simulations and test-particle theories for cosmic ray transport. I. Magnetostatic turbulence. J. Phys. G **32**, 809 (2006a)

Tautz, R.C., Shalchi, A., Schlickeiser, R.: Comparison between test-particle simulations and test-particle theories for cosmic ray transport. II. Plasma wave turbulence. J. Phys. G **32**, 1045 (2006b)

Taylor, G.I.: Diffusion by continuous movement. Proc. Lond. Math. Soc. **20**, 196 (1922)

Teufel, A., Schlickeiser, R.: Analytic calculation of the parallel mean free path of heliospheric cosmic rays. I. Dynamical magnetic slab turbulence and random sweeping slab turbulence. Astron. Astrophys. **393**, 703 (2002)

Teufel, A., Schlickeiser, R.: Analytic calculation of the parallel mean free path of heliospheric cosmic rays. II. Dynamical magnetic slab turbulence and random sweeping slab turbulence with finite wave power at small wavenumbers. Astron. Astrophys. **397**, 15 (2003)

Teufel, A., Lerche, I., Schlickeiser, R.: Cosmic ray transport in anisotropic turbulence. II. Shear Alfvén waves. Astron. Astrophys. **397**, 777 (2003)

The Pierre Auger Collaboration: Correlation of the highest-energy cosmic rays with nearby extragalactic objects. Science, **318**, 938 (2007)

Tu, C.-Y., Marsch, E.: A model of solar wind fluctuations with two components – alfvén waves and convective structures. J. Geophys. Res. **98**, 1257 (1993)

Urch, I.H.: Charged particle transport in turbulent magnetic fields: the perpendicular diffusion coefficient. Astrophys. Space Sci. **46**, 389 (1977)

Urch, I.H.: Simulations of cosmic-ray particle diffusion. Astrophys. Space Sci. **104**, 357 (1984)

Völk, H.J.: Nonlinear perturbation theory for cosmic ray propagation in random magnetic fields. Astrophys. Space Sci. **25**, 471 (1973)

Völk, H.J.: Cosmic ray propagation in interplanetary space. Rev. Geophys. Space Phys. **13**, 547 (1975)

Völk, H.J., Zank, L.A., Zank, G.P.: Cosmic ray spectrum produced by supernova remnants with an upper limit on wave dissipation. Astron. Astrophys. **198**, 274 (1988)

Webb, G.M., Zank, G.P., Ko, C.M., Donohue, D.J.: Multidimensional Green's functions and the statistics of diffusive shock acceleration Astrophys. J. **453**, 178 (1995)

Webb, G.M., Zank, G.P., Kaghashvili, E.K., leRoux, J.A.: Compound and perpendicular diffusion of cosmic rays and random walk of the field lines. I. Parallel particle transport models. Astrophys. J. **651**, 211 (2006)

Weinstock, J.: Formulation of a statistical theory of strong plasma turbulence. Phys. Fluids **12**, 1045 (1969)

Weinstock, J.: Turbulent plasmas in a magnetic field-a statistical theory. Phys. Fluids **13**, 2308 (1970)

Zank, G.P., Pauls, H.L., Cairns, I.H., Webb, G.M.: Interstellar pick-up ions and perpendicular shocks: implications for the termination shock and interplanetary shocks. J. Geophys. Res. **101**, 457 (1996)

Zank, G.P., Rice, W.K.M., Wu, C.C.: Particle acceleration and coronal mass ejection drive shocks: a theoretical model. J. Geophys. Res. **105**, 25079 (2000)

Zank, G.P., Li, G., Florinski, V., Matthaeus, W.H., Webb, G.M., le Roux, J.A.: Perpendicular diffusion coefficient for charged particles of arbitrary energy. J. Geophys. Res. **109**, A04107 (2004)

Zank, G.P., Li, G., Florinski, V., Hu, Q., Lario, D., Smith, C.W.: Particle acceleration at perpendicular shock waves: model and observations. J. Geophys. Res. **111**, A06108 (2006)

Zhang, M., Jokipii, J.R., McKibben, R.B.: Perpendicular transport of solar energetic particles in heliospheric magnetic fields. Astrophys. J. **595**, 493 (2003)

Zhou, Y., Matthaeus, W.H., Dmitruk, P.: Colloquium: magnetohydrodynamic turbulence and time scales in astrophysical and space plasmas. Rev. Mod. Phys. **76**, 1015 (2004)

Zimbardo, G., Pommois, P., Veltri, P.: Superdiffusive and subdiffusive transport of energetic particles in solar wind anisotropic magnetic turbulence. Astrophys. J. **639**, L91 (2006)
Zwickl, R.D., Webber, W.R.: Solar particle propagation from 1–5 AU. Solar Phys. **54**, 457 (1977)
Zwickl, R.D., Webber, W.R.: The interplanetary scattering mean free path from 1 to 3 × 1000 MV. J. Geophys. Res. **83**, 1157 (1978)

Index

Recently Published in the ASSL series

Volume 362: *Nonlinear Cosmic Ray Diffusion Theories*,
by Andreas Shalchi. ISBN: 978-3-642-00308-0

Volume 361: *The Sun Recorded Through History: Scientific Data Extracted from Historical Documents,* by J.M. Vaquero, M. Vázquez.
Hardbound ISBN: 978-0-387-92789-3

Volume 360: *The Principles of Astronomical Telescope Design,*
by Jingquan Cheng. Hardbound ISBN: 978-0-387-88790-6, February 2009

Volume 359: *Physics of Relativistic Objects in Compact Binaries: from Birth to Coalescence,* by M. Colpi, P. Casella, V. Gorini, U. Moschella, A. Possenti.
Hardbound ISBN: 978-1-4020-9263-3

Volume 358: *Cosmic Rays in Magnetospheres of the Earth and other Planets,*
by Dorman, Lev. Hardbound 978-1-4020-9238-1

Volume 357: *Neutron Stars and Pulsars,* by Becker, Werner.
Hardbound 978-3-540-76964-4

Volume 355: *Black Hole Gravitohydromagnetics,* by Punsly, Brian.
Hardbound 978-3-540-76955-2

Volume 354: *Sirius Matters*, by Noah Brosch. 978-1-4020-8318-1, March 2008

Volume 353: *Hydromagnetic Waves in the Magnetosphere and the Ionosphere*,
by Leonid S. Alperovich, Evgeny N. Fedorov. Hardbound 978-1-4020-6636-8

Volume 352: *Short-Period Binary Stars: Observations, Analyses, and Results*,
edited by Eugene F. Milone, Denis A. Leahy, David W. Hobill.
Hardbound ISBN: 978-1-4020-6543-9, September 2007

Volume 351: *High Time Resolution Astrophysics*, edited by Don Phelan, Oliver Ryan, Andrew Shearer. Hardbound ISBN: 978-1-4020-6517-0, September 2007

Volume 350: *Hipparcos, the New Reduction of the Raw Data*, by Floor van Leeuwen.
Hardbound ISBN: 978-1-4020-6341-1, August 2007

Volume 349: *Lasers, Clocks and Drag-Free Control: Exploration of Relativistic Gravity in Space*, edited by Hansjorg Dittus, Claus Lammerzahl, Salva Turyshev.
Hardbound ISBN: 978-3-540-34376-9, September 2007

Volume 348: *The Paraboloidal Reflector Antenna in Radio Astronomy and Communication Theory and Practice*, by Jacob W. M. Baars
Hardbound 978-0-387-69733-8, July 2007

Volume 347: *The Sun and Spae Weather*, by Arnold Hanslmeier.
Hardbound 978-1-4020-5603-1, July 2007

Volume 346: *Exploring the Secrets of the Aurora*, by Syun-Ichi Akasofu.
Hardboun 978-0-387-45094-0, July 2007

Volume 345: *Canonical Perturbation Theories Degenerate Systems and Resonance*,
by Sylvio Ferraz-Mello. Hardbound 978-0-387-38900-4, January 2007

Volume 344: *Space Weather: Research Toward Applications in Europe*, edited by
Jean Lilensten. Hardbound 1-4020-5445-9, January 2007

Volume 343: *Organizations and Strategies in Astronomy: Volume 7*, edited by
A. Heck. Hardbound 1-4020-5300-2, December 2006

Volume 342: *The Atrophysics of Emission Line Stars*, by Tomokazu Kogure,
Kam-Ching Leung. Hardbound ISBN: 0-387-34500-0, June 2007

For other titles see www.springer.com/astronomy

GPSR Compliance

The European Union's (EU) General Product Safety Regulation (GPSR) is a set of rules that requires consumer products to be safe and our obligations to ensure this.

If you have any concerns about our products, you can contact us on ProductSafety@springernature.com

In case Publisher is established outside the EU, the EU authorized representative is:

Springer Nature Customer Service Center GmbH
Europaplatz 3
69115 Heidelberg, Germany

Batch number: 10620141

Printed by Printforce, the Netherlands